Sperm Morphology of Domestic Animals

Sperm Morphology of Domestic Animals

J.H. Koziol, DVM, MS, DACT
Associate Professor of Food Animal Medicine and Surgery
Texas Tech University School of Veterinary Medicine
Amarillo, TX, USA

C.L. Armstrong, DVM, MS, DACT
Associate Veterinarian
Elgin Veterinary Hospital
Elgin, TX, USA

Registered Office
John Wiley & Sons, Inc., 111 River Street, Hoboken, NJ 07030, USA

Editorial Office
111 River Street, Hoboken, NJ 07030, USA

For details of our global editorial offices, customer services, and more information about Wiley products visit us at www.wiley.com.

Wiley also publishes its books in a variety of electronic formats and by print-on-demand. Some content that appears in standard print versions of this book may not be available in other formats.

Library of Congress Cataloging-in-Publication Data applied for

ISBN: 9781119769767 (hardback)

Cover Design: Wiley
Cover Image: Courtesy of J.H. Koziol, Westend61/Getty Images

Set in 9.5/12.5pt STIXTwoText by Straive, Pondicherry, India

Printed in Singapore
M114045_221221

For RLC and DFW who taught us to never be satisfied and complacent with the status quo

Contents

Preface

Sperm Morphology of Domestic Animals has been written to further the understanding of morphologic assessment in the dog, stallion, bull, buck, and ram. This text is directed toward veterinary students and practitioners. Morphologic evaluation of sperm is a confusing part of the fertility examination. Ample images and figures have been added to provide real-time support during semen evaluation. The authors sincerely hope the added imagery will help provide clarity to a complex and often confusing portion of fertility assessment. Morphologic evaluation of sperm is an ever-evolving area of theriogenology, and this text has been written with the latest information from the current literature.

The authors would like to thank their mentors and other leaders in the field of theriogenology that inspire excellence in all that they set out to accomplish. This text would not have been possible without the generous support from Wiley.

About the Companion Website

This book is accompanied by a companion website:

www.wiley.com/go/koziol/sperm

The website includes:

- Teaching PowerPoints

Introduction

Sperm evaluation provides a noninvasive method to evaluate testicular and epididymal function, providing information similar to that gained by a testicular biopsy. An abnormal spermiogram with supporting evidence from the history and physical exam can give insights into reasons for abnormal testicular function, and consequently allow formation of a prognosis for recovery or potential treatment. When an abnormal spermiogram is found, the types and number of abnormalities combined with history regarding environment, nutrition, and health status can be used to compile a reason for spermiogram disturbances noted. The veterinarian can then use that information to make a diagnosis and prognosis for recovery.

An awareness of the mechanism of sperm transport in the female reproductive tract aids in understanding the relationship between sperm quality and fertility and consequently the level of defects that can be tolerated and still maintain reasonable pregnancy rates. The cervix, uterus, and uterotubal junction all reduce the number of abnormal sperm that can reach the site of fertilization. The cervix filters sperm with tail defects, while severely deformed heads are filtered at the level of the cervix, uterus, and uterine tubes. However, the filter system is not perfect and some sperm abnormalities may still reach the oocyte and induce fertilization [1, 2].

Three main criteria should be taken into account when evaluating abnormal sperm morphology: (i) abnormalities of the nucleus that allow ovum penetration and zona reaction but not fertilization or embryonic development *cannot* be tolerated at levels >15–20% without subsequent decrease in pregnancy rate; (ii) abnormalities of the acrosome and sperm tail do not interfere with the ability of normal sperm to fertilize ova and therefore can be tolerated at levels up to 25%; (iii) a least 70–75% of spermatozoa should be normal [3]. The types and percentages of defects should be taken into consideration in both natural and artificial mating systems.

I.1 Spermatogenesis

A thorough understanding of normal and abnormal spermatogenesis facilitates understanding and interpretation of a spermiogram (description of sperm morphology noted during evaluation) (Figures I.1–I.4). The testis is largely composed of seminiferous tubules and interstitial tissue. The latter is located between the seminiferous tubules and consists primarily of Leydig cells, which produce and secrete steroid hormones, as well as vascular and lymph vessels that supply the testicular parenchyma. Seminiferous tubules arise from primary sex cords and contain germinal cells, as well as Sertoli cells, which support and nurture production of sperm. Sertoli cells form tight junctions,

Sperm Morphology of Domestic Animals, First Edition. J.H. Koziol and C.L. Armstrong.
© 2022 John Wiley & Sons, Inc. Published 2022 by John Wiley & Sons, Inc.
Companion website: www.wiley.com/go/koziol/sperm

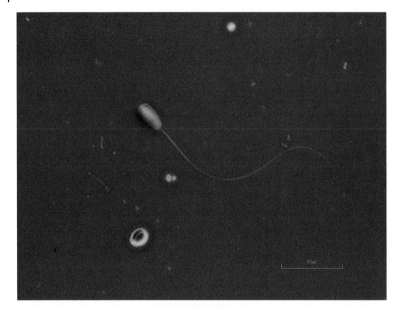

Figure I.1   Morphologically normal sperm from a bull (eosin–nigrosin, 1000×).

creating the most important portion of the blood–testis barrier [4]. Spermatogenesis occurs primarily in the tubulus contortus section of the seminiferous tubules. These tubules are surrounded by contractile peritubular cells that promote flow out of the tubulus contortus into the rectus (straight portion) of the seminiferous tubule. Both ends of each tubule are connected with the rete testis.

Spermatogenesis can be divided into three phases. The first phase, proliferation, consists of six mitotic divisions of spermatogonia, increasing the number of A-spermatogonia then B-spermatogonia. An important part of this phase is stem cell renewal. Loss of intercellular bridges allow some spermatogonia to revert to stem cells. B-spermatogonia then undergo several mitotic divisions, with the last division resulting in primary spermatocytes. The meiotic phase begins with diploid primary spermatocytes. During meiosis I, genetic diversity is ensured by DNA replication and crossing over during production of secondary spermatocytes, and as a result, from a genetic perspective, no two sperm are identical. The last phase, known as the differentiation phase or spermiogenesis, is differentiation of a spherical undifferentiated spermatid into a fully differentiated sperm with a head, a flagellum which includes the midpiece, and the principal piece. The length of spermiogenesis differs between species. An example is that it takes the last 18 days for the spermiogenesis stage of spermatogenesis to be completed in the bull.

Correct clinical interpretation of spermiograms also requires understanding of the specific timing of spermatogenesis in the species being evaluated. An understanding of the cycle of the seminiferous epithelium aids in this mission. The cycle of seminiferous epithelium is the progression through a complete series of stages at one location along a seminiferous tubule. On cross-sectional evaluation of the seminiferous tubule, four to five concentric layers of germ cells are present with each layer representing a generation. Each cross-section along the length of the seminiferous tubule will have a distinct appearance. Each cross section with its four to five generations of germ cells represents a stage of the seminiferous epithelium cycle. For example, a cross section in stage 1 will have a base layer of A-spermatogonium followed by primary spermatocytes, and another layer of primary spermatocytes, followed by a layer of immature spermatids. A stage 8 cross section will begin with a layer of A-spermatogonium, followed by B-spermatogonium, primary

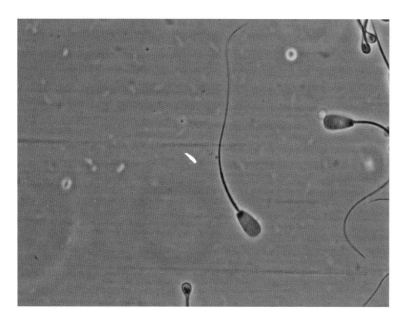

Figure I.2 Morphologically normal sperm from a bull (phase contrast wet mount, 1000×).

spermatocytes, immature spermatids, and mature spermatids that are ready to be released into the lumen. Along the length of any given seminiferous tubule, there are many zones or cross sections in different stages with only a few zones at an appropriate stage to release mature spermatid into the lumen of the seminiferous tubule, which will then travel through the rete testis to be transported to the head of the epididymis. These different zones allow for the creation of a spermatogenic wave or constant release of sperm along the length of each tubule. With zones in multiple

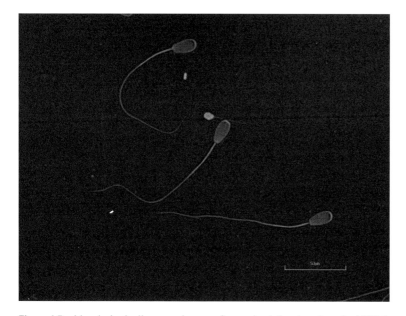

Figure I.3 Morphologically normal sperm from a buck (eosin–nigrosin, 1000×).

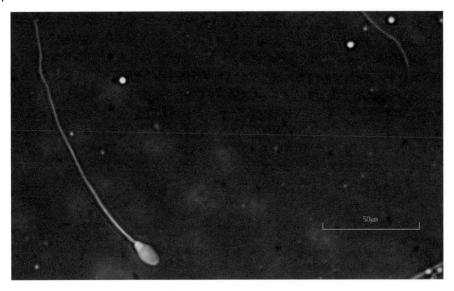

Figure I.4   Morphologically normal sperm from a stallion (eosin–nigrosin, 1000×).

stages along the length of the tubule, there is a constant flow of spermiation, which allows for a constant flow of sperm to the epididymis.

Normal testicular function is dependent on appropriate endocrine, autocrine, and paracrine control. In the male, gonadotropin releasing hormone (GnRH) is secreted from the hypothalamus in a pulsatile nature resulting in waves of luteinizing hormone (LH) and follicle stimulating hormone (FSH) release. LH acts on the Leydig cells within the testes, which are responsible for synthesizing progesterone, the majority of which is converted to testosterone. This pulsatile nature of LH is necessary for normal testicular function as sustained LH can lead to a refractory nature of the Leydig cells secondary to a downregulation in the number of receptors. This culminates in reduced secretion of testosterone from the Leydig cells. High concentrations of testosterone must be maintained for normal spermatogenesis to occur. The pulsatile release of LH and consequently testosterone prevents negative inhibition of FSH, which is needed for maintenance of Sertoli cells. Sertoli cells convert testosterone to dihydrotestosterone and estradiol along with secreting androgen binding protein (ABP). Approximately 80% of the secreted ABP goes into the lumen of the seminiferous tubule and travels to the epididymis, while the remainder is secreted into the interstitial compartment and is absorbed into the systemic circulation. ABP binds testosterone and is responsible for not only maintaining a high concentration of testosterone within testes but also carries testosterone to the epididymis.

The most common causes of abnormal spermatogenesis in males include abnormal testicular thermoregulation; hormonal imbalances, particularly those associated with stress; and effect(s) of toxins or expression of deleterious genes [5]. Stress typically elevates systemic cortisol concentrations, profoundly decreasing release of LH and testosterone [6–8]. Stress has many origins, including environment, illness, or injury, causing changes in the spermiogram similar to those induced by disruption of thermoregulation. The primary spermatocyte is extremely sensitive to alterations in the hormonal milieu secondary to stress or illness. For example, in cases of testicular degeneration, the changes appear first in the cytoplasm, centrosomes, and spindles at the level of the primary spermatocyte, which predispose to disturbances in the developing spermatid [3, 9].

I.2   Length of Spermatogenesis in Domestic Species

Stage	Bull [10]	Ram [10]	Boar [10]	Stallion [10]	Dog [11]
I	4.2	2.2	1.1	2.0	3.0
II	1.2	1.1	1.4	1.8	1.7
III	2.7	1.9	0.4	0.4	0.4
IV	1.7	1.1	1.2	1.9	1.6
V	0.2	0.4	0.8	0.9	1.1
VI	0.8	1.3	1.6	1.7	2.1
VII	1.1	1.1	1.0	1.6	1.8
VIII	1.6	1.0	0.8	1.9	1.9
Length of one cycle	13.5	10.4	8.3	12.2	13.6
Total days for spermatogenesis	61	47	39	57	61

I.3   Maturation of Spermatozoa in the Epididymis

From the rete testis, sperm travel to the efferent duct and into the head of the epididymis. Grossly, the epididymis can be divided into three segments: the head (caput), body (corpus), and tail (cauda). The epididymis is responsible for transport, concentration, storage, and maturation of sperm.

During epididymal transport, maturation of sperm continues with many biochemical and morphological modifications occurring. All sperm that enter the head of the epididymis do so with a proximal cytoplasmic droplet; as they move through the epididymis, the droplet normally migrates down the midpiece and is present at the distal end of the midpiece when sperm arrive in the tail of the epididymis.

Along the entire length of epididymis, the lumen is bordered by epithelium that synthesizes and secretes proteins under androgen stimulation [12]. Intraluminal fluids released sequentially along the epididymal lumen are key for maturation of sperm [13]. As the sperm cell travels along the lumen many biochemical modifications occur. These processes include nucleus chromatin condensation, changes in plasma membrane composition and relocation of surface antigens, and the ability to respond to hypo-osmotic stress [14]. Circulating and intraluminal androgens are critical for epididymal gene expression and secretion. Androgens in the epididymis are synthesized by enzymatic reduction of testicular testosterone. Estrogen produced by the testicular Leydig cells also plays a role in the epididymis particularly in the role of water reduction in the proximal regions of the epididymis [14].

Following ejaculation, distal droplets should be shed after mixing with seminal fluid. Epididymal transport time is species dependent (Table I.1). The tail of the epididymis and ductus deferens serve as short-term sperm storage. However, in the absence of ejaculation, senescent sperm should be continually expelled into the urinary tract, ensuring that viable sperm are always available, even following long periods of sexual rest [15].

I.4   Staining Techniques and Evaluation of Morphology

Microscopic evaluation of sperm morphology is a pillar of successful sperm evaluation and breeding soundness examinations of any species. The three basic types of light microscopy are brightfield, phase contrast, and differential interface contrast (DIC). Most conventional microscopes have a brightfield function and many also have phase contrast capability. However, DIC is expensive and is generally restricted to use in research or andrology laboratories.

Brightfield microscopy is suitable for examination of morphology utilizing stained semen smears. Morphology evaluations must always be performed using the oil immersion (100×) objective lens; with eyepieces of 10–12.5×, the total magnification is 1000–1250× (hereafter referred to as 1000×). Morphology evaluation at lower magnification is not recommended as it results in failure to recognize many sperm abnormalities.

Utilization of phase contrast microscopy allows examination of unstained, cover-slipped, preparations prepared by mixing semen with warmed 10% neutral buffered formalin on a microscope slide, as well as slides stained with Feulgen stain [20]. Evaluation of sperm morphology utilizing phase contrast at 1000× is preferred over the use of stained preparations by some practitioners, as defects such as nuclear vacuoles, [21] acrosomal defects, and abnormal DNA in the case of Feulgen-stained slides are easier to identify. However, these defects can also readily be identified with high-quality brightfield microscopes.

Morphologically abnormal sperm have long been associated with male infertility, and assessment of these abnormalities is a fundamental component of analysis of semen quality. [3, 22–24]. Sperm morphology is an excellent predictor of the outcome of natural mating, artificial insemination, and in vitro fertilization, [3, 25, 26] and there is a relationship between morphologically abnormal sperm and poor DNA quality [24].

I.5   Vital Staining

Live/dead stains, also referred to as supravital stains, such as eosin–nigrosin and eosin–aniline blue, are commonly used for routine morphological examination of sperm [27]. In a study by Tanghe et al., eosin–nigrosin staining of sperm was the only bull sperm quality parameter evaluated that demonstrated a significant association with pronucleus formation in in vitro fertilization [28]. When utilizing live/dead stains, nigrosin or aniline blue provides a background, whereas eosin acts as a vital stain. Functional plasma membranes of viable sperm prevent penetration of eosin into the cell; therefore, viable sperm appear white. In contrast, eosin penetrates damaged cell membranes, staining injured or nonviable sperm pink.

Table I.1   Mean time of sperm epididymal passage in domestic

Species	Average time (days)	References
Dog	15	Linde-Forsberg [16]
Boar	9–14	Franca and Cardoso [17]
Bull	8–11	Barth and Oko [3]
Stallion	8–11	Swierstra [18]
Ram	16.4	Amann [19]

Figure I.5 Partially stained sperm with a pink posterior and a white anterior (eosin–nigrosin, bull, 1000×).

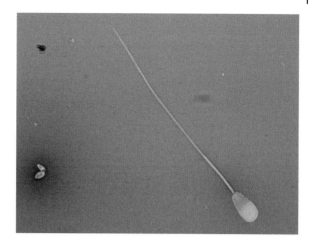

   Partially stained sperm, with a pink posterior portion and a white anterior portion of the sperm head, are sometimes observed. These sperm have suffered membrane damage, yet the acrosome remained intact. This staining pattern occurs because the acrosomal membrane is tightly applied to the nucleus at the equatorial region and penetration of the eosin stain under the acrosome is initially impeded at the equatorial region [29]. Partially stained sperm are more common when semen samples are processed under less than ideal conditions, suggesting that sperm membrane damage or cell death occurred very recently (Figure I.5).

   When assessing the proportion of sperm staining alive, it may be relevant to only count sperm stained completely white. When semen has been handled properly and stained correctly, the percentage excluding eosin, and therefore considered "alive" is highly correlated to progressive motility. **This is not to suggest that only the "alive" cells be counted during evaluation of a sperm morphology slide; 100 cells should be counted regardless of vital status to give an accurate representation of the current state of the ejaculate. The accuracy of the assessment of marginal bulls can be improved by counting 300 cells.**

I.6   Preparation of Semen Smears or Coverslip Slides

Slide preparation is crucial for proper evaluation of sperm morphology and requires practice and attention to detail. Common errors include too many sperm per slide (Figure I.6); slides stained too dark or too light; and slow drying resulting in cracks and stain artifacts. There are multiple ways to create stained slides. Some may prefer to streak out slides using another glass slide in a manner similar to producing a blood smear while others prefer to use a wood applicator stick to roll the sample out (Figures I.7–I.9). Some practitioners believe that the wood applicator stick results in less mechanical damage to the sperm (detached heads, damaged acrosomes). However, no studies have proven this clinical suspicion. Slides can be viewed anywhere along the length but usually half way down the slide is a good place to start as the cells are usually evenly distributed with good background stain.

   Wet mounts can be made for evaluation of sperm morphology by phase-contrast or DIC microscopy at 1000×. Freshly ejaculated semen must be diluted and motility stopped to allow for accurate evaluate. This can be accomplished with the addition of 0.2% glutaraldehyde in phosphate buffered saline (PBS), buffered formal saline, or with neutral buffered formalin. To prepare a wet mount slide, a 2–3 mm drop of diluted semen is placed on a warm clean microscope slide and a coverslip is placed over top. Allow the drop to evenly distribute under the coverslip before placing a drop of oil

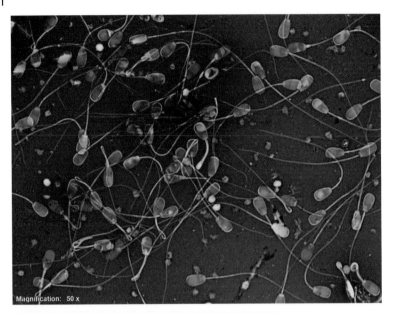

Figure I.6 Poorly prepared slide with too many sperm per field, which impeded the evaluators ability to evaluate each individual sperm (eosin–nigrosin, 500×).

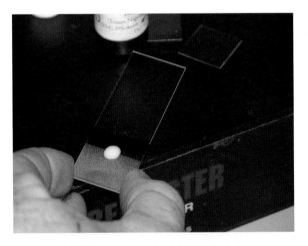

Figure I.7 A small line of stain is placed along the slide followed by a small drop of semen.

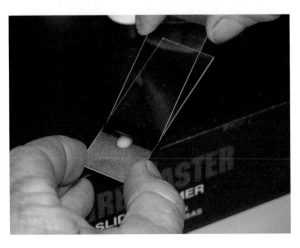

Figure I.8 Using a second slide, one can start in the stain backing up to pick up a small line of semen and then proceed to streak out the slide. There are many different methods than can be used for streaking out slides.

Figure I.9 Slides should neither be too dark or too light, and when viewed under the microscope, the cells should be spaced in such a manner that they can each be individually examined.

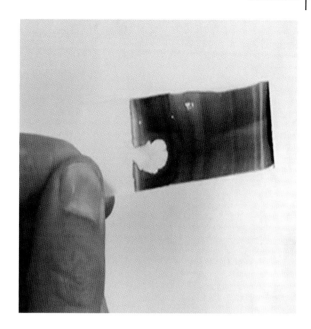

on the coverslip for examination at 1000×. The preparation should be very thin to avoid sperm laying on their edges rather than flat. When using glutaraldehyde or neutral-buffered formalin, clumping of cells may occur. Wet mount slides can be viewed at any location along the coverslip with the exception of near the edges, as this is where a higher number of dead cells tend to accumulate [3].

I.7 Differential Counting of Sperm Morphology

Differential counts of sperm morphology must be performed using 1000× magnification (oil immersion) as lower magnifications result in failure to recognize some defects. The goal for examining sperm morphology is to determine the percentage and types of sperm abnormalities present in a sample and thereby construct a spermiogram. The spermiogram may be used to interpret potential fertility of semen at the time of sampling, as well as determine potential causes of an abnormal spermiogram. Knowledge of probable cause of abnormalities may allow the practitioner to develop a prognosis for recovery, which will aid the client in making an informed decision about the male in question [29].

Determination of a spermiogram is facilitated by the use of a differential cell counter. Mechanical cell counters often allow one to press two or more keys simultaneously to record more than one defect within the same sperm, while counting the individual sperm only once. With this system, a more accurate spermiogram may be produced.

When sperm morphology is good, differential counting of 100 sperm is adequate to determine a spermiogram. However, when the percentage of abnormal sperm present is high, counting at least 300 sperm may be necessary to produce an accurate spermiogram.

Section I

Head Abnormalities

1

Pyriform and Tapered Heads

The prevalence of pyriform head defects are relatively high and are either the most or second most prevalent defect in an ejaculate [3, 30–34]. The classical pyriform sperm has a pear-shaped head, a normal acrosome, and a narrow post-acrosomal region [3, 35–37]. Sperm with irregularity in head shape and size generally have good motility [3].

Several variants of pyriform head defects have been reported, ranging from slightly affected to severely tapered through the acrosome and post-acrosomal regions. The main initiating factor for this deformity appears to be adverse environmental conditions impacting male health and fitness [3, 38]. Pyriform sperm have been noted in ejaculates three to four weeks after initiation of dexamethasone injections to simulate stress, following scrotal insulation, or exposure to high ambient temperatures in the case of rams [3, 5, 38]. Pyriform or tapered heads have also been noted in those suffering from obesity resulting in scrotal or inguinal deposition of fat, frostbite or dermatitis of the scrotum, varicoceles, or other injuries that impede proper testicular thermoregulation [3]. Disruptions in endocrine hormone balance occurring either systemically or locally may also be implicated. Local disruption is caused by abnormalities in testicular thermoregulation while systemic disruption descends from unfavorable conditions or events that cause stress to the animal. These stressful events can include chronic pain, joint pain, conditions affecting the hooves including lameness or laminitis, extremes in weather, housing or social stress, improper nutrition, or travel [3].

Differences in the severity of response between males with similar treatments suggest that some males may be genetically predisposed to develop pyriform heads in response to adverse environmental influences [3, 5, 29]. In most instances, there will be many other sperm abnormalities present in the ejaculate, indicating a disturbance in spermatogenesis has occurred. However, after removal of stress factors, males often return to normal sperm production [3].

Small numbers of pyriform or tapered defects can be found in the semen of most bulls, even in bulls of good fertility [3]. In a study of bulls used for natural service and artificial insemination, percentage of bulls producing affected sperm were 9 and 16%, respectively. The percentage of pyriform sperm in an ejaculate from affected individual varies from 10 to >50% [3]. When pyriform defects are in high numbers, their impact on fertility has been documented in both in vivo and in vitro studies. Thundathil et al. [39] reported that in vitro fertilization rates were lower for semen containing high numbers of pyriform heads than for ejaculates containing low numbers, 68.5 versus 84.4%, respectively. Sixty-day pregnancy rates in estrus-synchronized artificially inseminated heifers using semen from these same bulls were 37 and 61%, respectively, for bulls with pyriform sperm and the control bull. In the same study, rates of embryo/fetal loss between rates of embryo/fetal loss between days 22 and 60 of pregnancy were 23 and 8% for the pyriform head group and the control group, respectively [39].

Pyriform sperm disrupt various stages of the fertilization process including: sperm transport, [40–42] oocyte binding and zona penetration, [43] along with postfertilization events [39]. Furthermore, the prognosis for recovery is highly variable among animals. Mature bulls with high levels (25–75%) of pyriform sperm have a poor prognosis for recovery when there is no

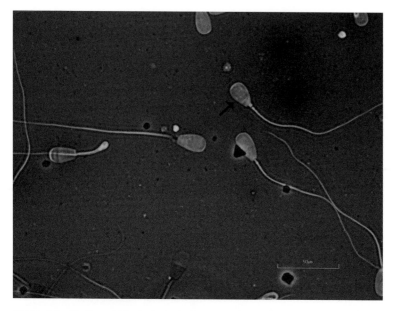

Figure 1.1 Pyriform head in a bull (eosin–nigrosin, 1000×).

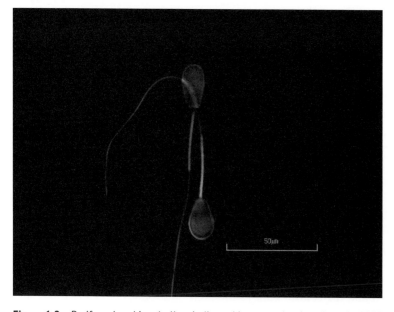

Figure 1.2 Pyriform head in a bull as indicated by arrow (eosin–nigrosin, 1000×).

apparent reason for a disturbance of spermatogenesis [21]. Return to normal sperm production by bulls known to have suffered an adverse condition predisposing them to disturbed spermatogenesis is dependent on the severity and duration of the insult. Younger over-conditioned bulls often produce an ejaculate with a high level of pyriform heads and recover after weight loss [3].

In stallions, Jasko and Love both noted a negative correlation between the percentage of sperm head defects and fertility [32, 33]. Jasko et al. noted that head defects accounted for the large proportion of variation in per cycle pregnancy rates [33]. Love et al. similarly noted that a 1% increase in percentage of head defects resulted in a 0.67% reduction in per cycle pregnancy rates.

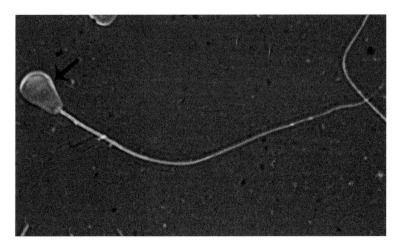

Figure 1.3 Pyriform head in a bull as indicated by arrow (eosin–nigrosin, 1000×).

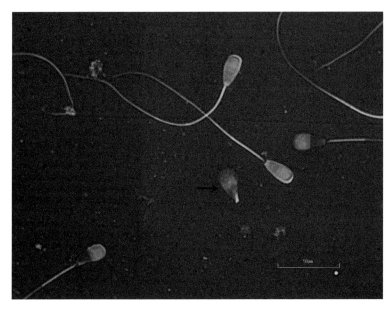

Figure 1.4 Detached pyriform head in a bull indicated by arrow (eosin–nigrosin, 1000×).

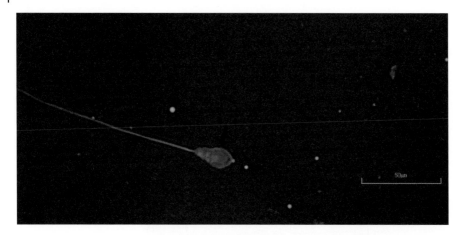

Figure 1.5 Pyriform head in bull as indicated by arrow (eosin–nigrosin, 1000×).

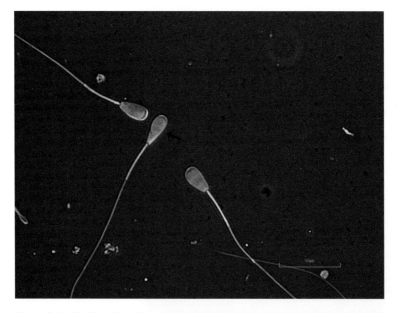

Figure 1.6 Pyriform head in a bull as indicated by arrow (eosin–nigrosin, 1000×).

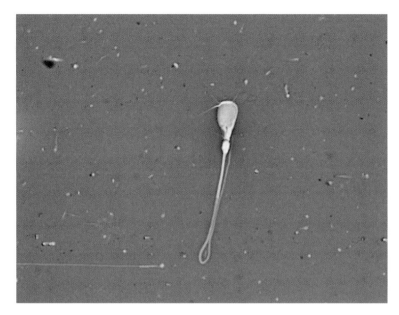

Figure 1.7 Pyriform head in bull as indicated by arrow also note the proximal droplet and terminally coiled tail (eosin–nigrosin, 1000×).

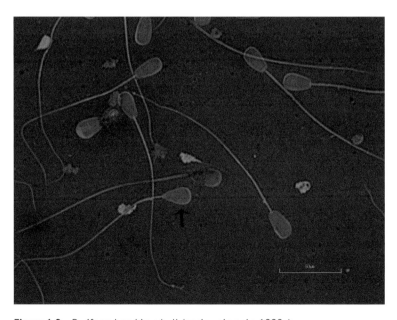

Figure 1.8 Pyriform head in a bull (eosin–nigrosin, 1000×).

Figure 1.9 Pyriform head in a ram (eosin–nigrosin, 1000×).

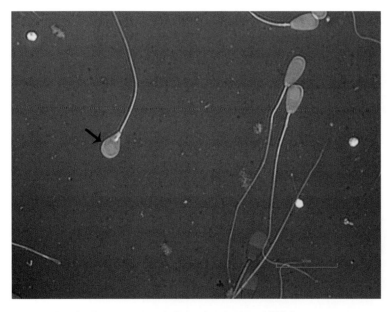

Figure 1.10 Pyriform head in a bull (eosin–nigrosin, 1000×).

Figure 1.11 Pyriform head defect in a bull as indicated by arrow compared to normal-shaped head as indicated by star (eosin–nigrosin, 1000×).

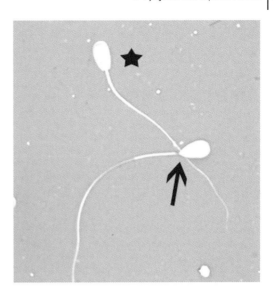

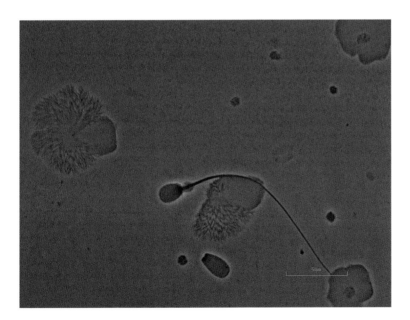

Figure 1.12 Pyriform-shaped head in a bull with proximal droplet (phase contrast wet mount, 1000×).

2

Nuclear Vacuolation, Including the Diadem Defect

Nuclear vacuoles have been described in sperm heads of all domestic species [44–50]. Historically, this defect has also been referred to as pouches or craters [49–52]. The defect is characterized by an invagination of the nuclear membrane into the nucleoplasm of the sperm head [53]. Nuclear vacuole formation has been identified as beginning as early as step 6 of spermiogenesis and has been identified in early and late spermatids. This coincides with the onset of condensation of the spermatid nucleoplasm and lengthening and flattening of the nucleus [54]. Vacuoles usually contain inclusions apparently derived from spermatid cytoplasm [29]. Various initiators of nuclear vacuolation have been postulated, including stress, aberrant testicular thermoregulation, viral infections, toxins, improper nutrition, and inheritance [29]. Dexamethasone treatments to mimic effects of stress in bulls acutely increased percentage of vacuoles, followed by a decline over several weeks [3, 5, 55]. Maximal levels of vacuoles occurred two to four weeks after dexamethasone treatment (20 mg/day for seven days given IM). In addition, scrotal insulation induced production of sperm head vacuolation in some bulls [5].

The incidence of nuclear vacuolation in the spermiogram of affected males can vary from <1% to nearly 100% [44]. When viewed using eosin–nigrosin morphology stain, vacuoles appear as dark spots in the nucleus and must be differentiated from artifacts such as stain crystals or other debris. Nuclear vacuoles are most easily spotted with phase contrast or differential interference microscopy. Feulgen staining will also reveal defects in the nuclear DNA [3]. Single or multiple vacuoles can be present anywhere in the sperm nucleus. However, they tend to be most commonly observed at the postnuclear cap-acrosome junction or near the apical ridge [55].

Nuclear vacuoles have three forms: single apical vacuoles present anywhere in the sperm head; large confluent vacuoles; or multiple pinpoint nuclear vacuoles at the acrosome post-acrosomal junction, resembling a string of pearls known as diadem vacuoles [44, 45, 56–58].

There are multiple reports that sperm with nuclear vacuoles impair fertility [44, 57, 59–61]. In one study, superovulated cows bred with semen from a bull with 80% nuclear vacuoles had a fertilization rate of 18% compared to that of 72% for control bulls [57]. Saacke demonstrated that sperm with nuclear vacuoles could transverse the female reproductive tract and access the ovum, thereby competing for fertilization [1]. However, vacuolated sperm were deficient in their ability to bind and penetrate the zona pellucida of *in vitro-matured* bovine oocytes [62]. High levels of nuclear vacuoles have also been shown to impair fertility in the stallion with semen from a stallion with 57% nuclear vacuoles producing no pregnancies over three years [63].

Sperm vacuoles have been noted to disappear in a few weeks following insult or persist for multiple years [3, 57, 61]. Collective results of available reports indicate a genetic predisposition to formation of nuclear vacuoles following disruptions in thermoregulation or stress; however, levels of vacuolated sperm can fluctuate dramatically over short intervals without apparent reason, and response to stressors is very unpredictable. Animals producing sperm severely affected by nuclear vacuoles should be closely monitored for the rest of their reproductive careers [29]. Considering recovery of the defect prognosis primarily depends on etiology of the problem.

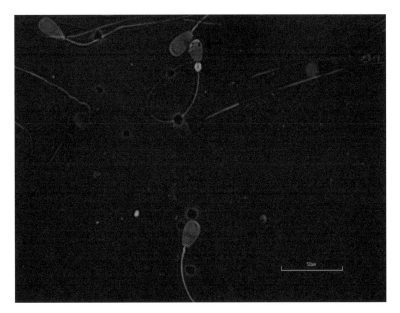

Figure 2.1 Nuclear vacuoles as indicated with arrow also note the proximal droplet (eosin–nigrosin, 1000×).

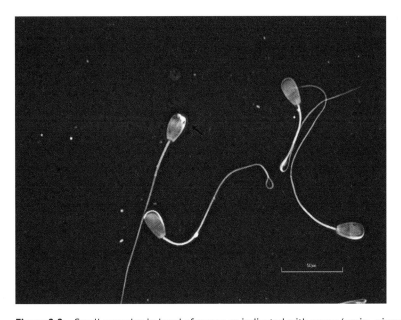

Figure 2.2 Small vacuoles in head of sperm as indicated with arrow (eosin–nigrosin, 1000×).

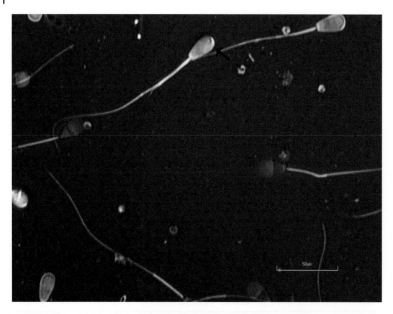

Figure 2.3   Vacuole in the head of sperm from a bull (eosin–nigrosin, 1000×).

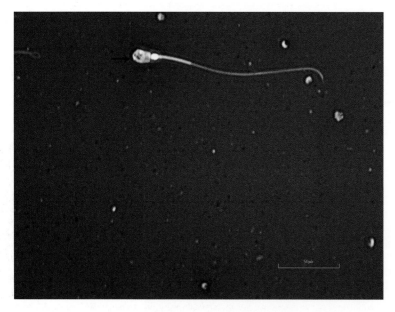

Figure 2.4   Nuclear vacuoles in a bull also note the proximal droplet (eosin–nigrosin, bull, 1000×).

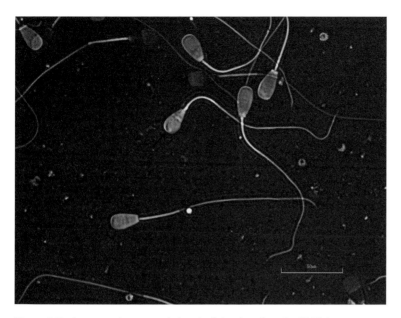

Figure 2.5 Large nuclear vacuole in a bull (eosin–nigrosin, 1000×).

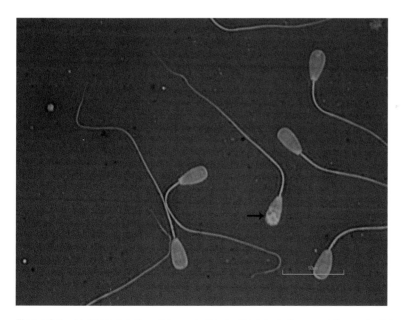

Figure 2.6 Multiple small nuclear vacuoles in the head of a sperm from a bull (eosin–nigrosin, 1000×).

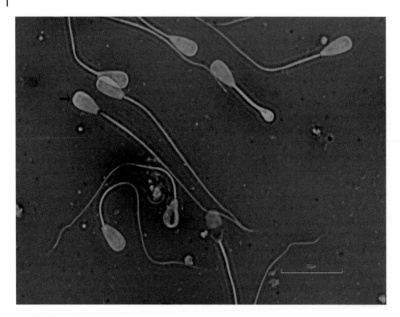

Figure 2.7 Multiple small nuclear vacuoles in the head of a sperm from a bull (eosin–nigrosin, 1000×).

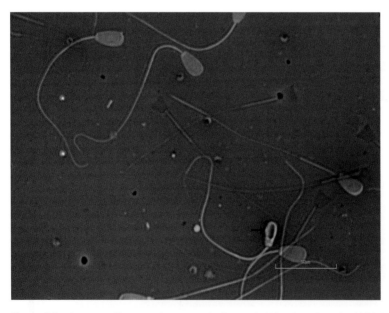

Figure 2.8 Large confluent nuclear vacuole from a bull (eosin–nigrosin, 1000×).

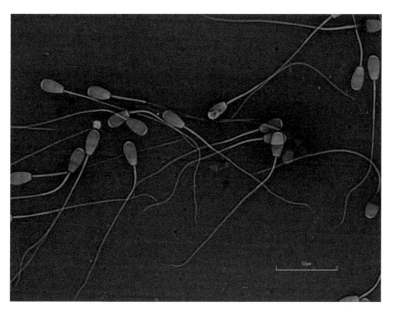

Figure 2.9   Multiple small nuclear vacuoles in a buck (eosin–nigrosin, 1000×).

Figure 2.10   Large confluent nuclear vacuoles in the head of sperm from a bull (eosin–nigrosin, 1000×).

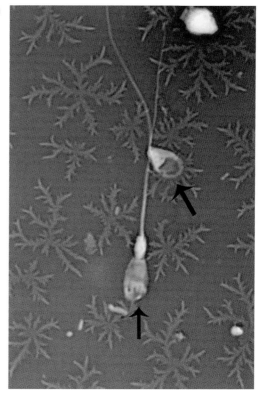

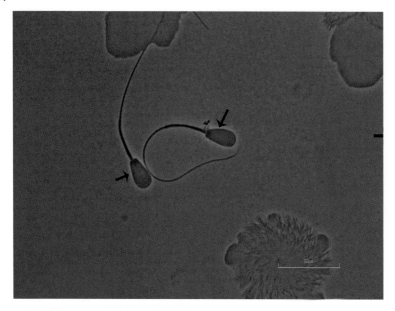

Figure 2.11 Diadem defects in a bull (phase contrast wet mount, 1000×).

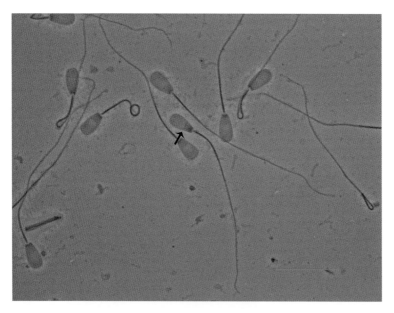

Figure 2.12 Diadem defect in a bull (phase contrast wet mount, 1000×).

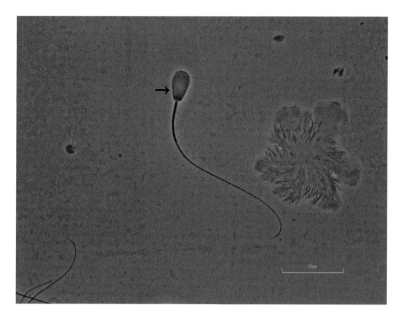

Figure 2.13 Diadem defect in a bull (phase contrast wet mount, 1000×).

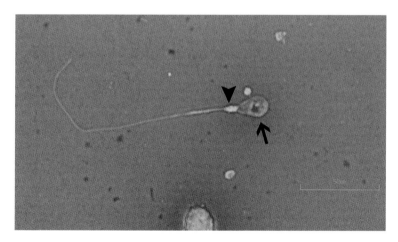

Figure 2.14 Large vacuole in the head (arrow). Also note the pseudodroplet (arrowhead) followed by mitochondrial aplasia (eosin–nigrosin, bull, 1000×).

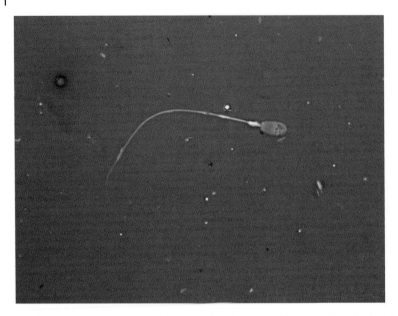

Figure 2.15 Multiple vacuole in the head of the sperm. Also note the mitochondrial disruption in the midpiece (eosin–nigrosin, bull, 1000×).

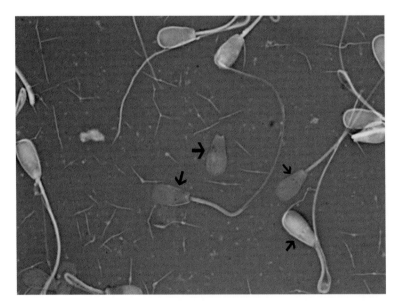

Figure 2.16 Multiple diadem defects indicated by arrow (eosin–nigrosin, bull, 1000×).

3

Macrocephalic and Microcephalic Sperm

Macrocephalic and microcephalic sperm are those for which the sperm head is clearly larger or smaller, respectively, than the normal population of sperm heads in a preparation. Although common in spermiograms, abnormally sized sperm heads are not often in large proportions, and there is minimal information in the literature regarding the origins of either macrocephalic or microcephalic sperm. Aberrant size of the sperm head may be due to an excess or deficiency of nuclear chromatin [9, 64]. Therefore, it is unlikely that macrocephalic and microcephalic sperm are able to participate in ovum fertilization and embryonic development [29].

The incidence of macrocephalic or microcephalic in spermiogram of animals with good fertility is nearly always less than 1% [3]. In cases of disrupted spermatogenesis demonstrated by the appearance of any number of head, midpiece, or tail defects, the incidence of macrocephalic or microcephalic heads will increase. Microcephalic heads are more likely to be observed in these cases than macrocephalic heads, and in general the percentage does not exceed 5–7% of the spermiogram [3].

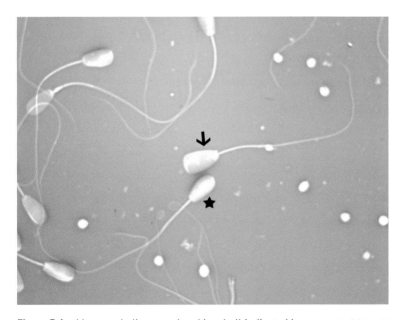

Figure 3.1 Macrocephalic sperm head in a bull indicated by arrow next to a normal sperm indicated by star (eosin–nigrosin, 1000×).

Sperm Morphology of Domestic Animals, First Edition. J.H. Koziol and C.L. Armstrong.
© 2022 John Wiley & Sons, Inc. Published 2022 by John Wiley & Sons, Inc.
Companion website: www.wiley.com/go/koziol/sperm

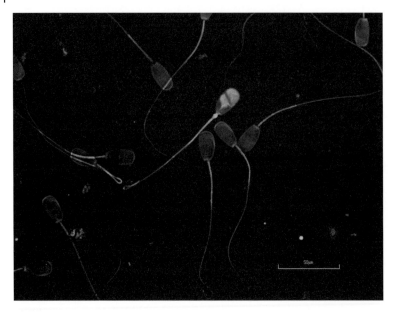

Figure 3.2 Macrocephalic head with proximal droplet in a bull (eosin–nigrosin, 1000×).

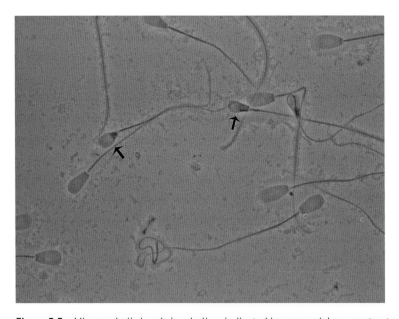

Figure 3.3 Microcephalic heads in a bull as indicated by arrows (phase contrast wet mount, 1000×).

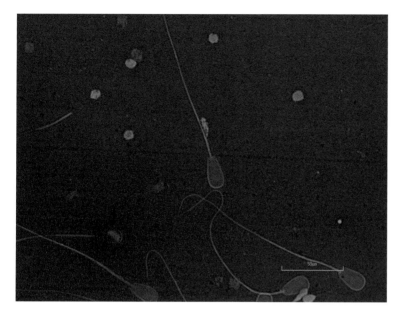

Figure 3.4 Microcephalic head in a bull (eosin–nigrosin, 1000×).

4

Rolled Head–Nuclear Crest–Giant Head Syndrome

Rolled heads, nuclear crests, or giant head syndrome are rare defects that have been reported in Brown Swiss, South Devon, and Friesian bulls and are thought to have a genetic basis [65–67]. These defects are found sporadically in normal and pathological semen samples in bulls [1]. The giant heads are often crested or rolled along the long axis to varying degrees and may be diploid or polyploid [38]. This defect has not been reported in other species.

Stress (environmental, social, or illness) along with toxicity may also result in the development of this abnormality [3].

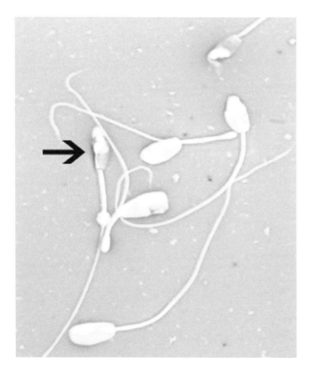

Figure 4.1 Rolled head (arrow): Rolled heads are often noted with the roll along the long axis of the head of the sperm (eosin–nigrosin, 1000×).

Sperm Morphology of Domestic Animals, First Edition. J.H. Koziol and C.L. Armstrong.
© 2022 John Wiley & Sons, Inc. Published 2022 by John Wiley & Sons, Inc.
Companion website: www.wiley.com/go/koziol/sperm

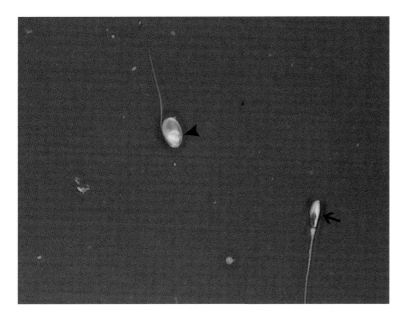

Figure 4.2 Rolled head in a bull indicated by arrow also note the teratoid indicated by the arrowhead (eosin–nigrosin, 1000×).

5

Abnormal DNA Condensation

In the early stages of spermatogenesis, sperm DNA is associated with large basic histone nucleo-proteins similar to other somatic cells. However, during the final stages of spermatogenesis, sperm chromatin structure is modified, and as spermiogenesis proceeds, histones are initially replaced by transition proteins that are subsequently replaced by small protamines [68]. This substitution of protamine allows DNA chains to lie parallel to each other to facilitate formation of a compact nucleus, resistant to denaturation. Incomplete condensation and partial retention of histones result in sperm malfunction [29].

Abnormal DNA condensation cannot be detected by standard light microscopic examination of unstained semen preparations or in routinely stained preparations that do not specifically identify chromosomal material or DNA. When infertility is observed in healthy animals with adequate semen quality, libido, and mating ability, abnormal DNA condensation may explain fertilization failure.

Sperm chromatin structure assays (SCSA) can be used to evaluate the character of chromatin in individual sperm [69]. SCSA involve acridine orange staining and assessment by flow cytometry to measure susceptibility of sperm chromatin to acid-induced denaturation [70].

Another, more practical way to assess DNA condensation is Feulgen staining, which allows visual microscopic detection of abnormal DNA condensation [3]. In the Feulgen reaction, aldehyde groups in the deoxyribose component of DNA are unmasked by acid hydrolysis and then exposed to Schiff's reagent [29]. Normal sperm nuclei appear as a homogenous purplish red with bright-field microscopy and intensely violet with phase microscopy at 1000×. In contrast, abnormal DNA condensation usually appears as coarse or fine clumping of nuclear material with a generalized reduced staining intensity of the nucleus. Affected nuclei often have an expanded appearance. In some cases, nuclei appear slightly mottled and pale. Nuclear vacuolation may be associated with abnormal sperm condensation. Concurrent control samples should always be used to confirm abnormal DNA staining with the Feulgen and other staining techniques [29]. The existence of clumped granular chromatin has also been detected with electron microscopy. Feulgen and SCSA methods have been compared and the two methods have good correlation with proportion of affected sperm and with fertility [71].

In normal bulls, the number of sperm with abnormal DNA condensation is <5%. A few bulls with >80% normal sperm when evaluated on an eosin-nigrosin-stained smear had 100% of nuclei with abnormal chromatin condensation when a Feulgen-stained smear was evaluated [29]. No breed predisposition has been noted [29]. Similarly, abnormal DNA condensation has been reported in the stallion, ram, boar, and dog [72–74].

Sperm Morphology of Domestic Animals, First Edition. J.H. Koziol and C.L. Armstrong.
© 2022 John Wiley & Sons, Inc. Published 2022 by John Wiley & Sons, Inc.
Companion website: www.wiley.com/go/koziol/sperm

Figure 5.1 Abnormal chromatin condensation in a bull as shown by arrows (Feulgen, 1000×).

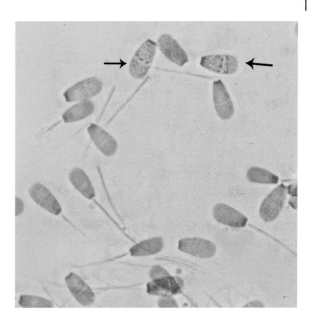

6

Acrosome Abnormalities

The acrosome contains multiple enzymes that promote fertilization by degradation of the outer membrane of the zona pellucida of the oocyte, thereby enabling sperm penetration to occur. For optimal fertility, sperm must have a normal acrosome characterized as a "tightly adherent, intact acrosome with a smooth surface and a distinct, uniformly shaped apical ridge" [75].

Structural aberrations including an incomplete or irregular apical ridge; an irregular acrosomal surface; and grossly distorted acrosomes including swollen, ruffled, or vacuolated forms are classified as acrosomal abnormalities [75]. All three types of acrosome aberrations have been associated with infertility in bulls, boars, rams, dogs, and stallions [3, 76–81].

The knobbed acrosome defect has been classically described as a bead-like extension at the apex of the affected sperm or a refractile or dark staining area of eccentric thickening at the apex of affected sperm [3, 82]. However, there is a wide variation in acrosomal bending that can be associated with the defect [77]. The more common appearance of the acrosomal defect is either flattened or indented sperm apex. In either form, the abnormal acrosome folds back on the sperm apex, yet only a few affected sperm have a bead-like protrusion at the sperm apex [3]. Stress, abnormal thermoregulation of the testes, or genetic causes have been postulated to result in acrosomal defects. Specific causes leading to flattened or indented acrosomes remain unknown. The concurrent appearance of other sperm abnormalities with acrosomal abnormalities likely indicates an adverse event such as stress or injury as the precipitating cause. However, a genetic predisposition should be considered when a high proportion of the sperm have acrosomal abnormalities and the defect persists in the absence of other sperm defects [3, 80, 83–85].

Sperm with knobbed acrosomes had reduced to no ability to bind to the zona pellucida; furthermore, apparently otherwise normal sperm from the same ejaculate also appeared to be functionally deficient. Oocytes penetrated by either knobbed sperm or morphologically normal sperm in the same ejaculate had reduced chances of fertilization, and resulting zygotes had reduced ability for cleavage and embryonic development to the blastocyst stage, suggesting this defect should be considered noncompensable [85]. Furthermore, sperm with acrosome defects have impaired plasma membrane function that predisposes them to premature capacitation and a spontaneous acrosome reaction [86, 87].

Another study presented four bulls with high numbers of the indented form of knobbed acrosomes were capable of fertilizing ova and supporting embryonic development at levels comparable to control bulls when used in single-sire mating systems [88]. However, when an affected bull that had been successful in a single-sire mating system was placed into a competitive mating trial with normal control bulls, he sired a significantly lower proportion of calves, despite actively competing for cows. Meyer postulated that based on this study [88] and others [82] that sperm with flattened or indented acrosomes are less competitive in fertilization than sperm with normal acrosomes [88].

In stallions, the presence of knobbed acrosomes has been associated with decreased fertility, but the degree to which it impacts fertility is confounded by factors including limited numbers of mares, and differences in breeding management of the mare [32, 77].

Figure 6.1 Acrosome abnormalities. Variation in appearance of knobbed acrosome defect. K-beaded form of the knobbed acrosome; I – indented form of the knobbed acrosome with arrow pointing toward indentation (eosin–nigrosin, bull, 1000×).

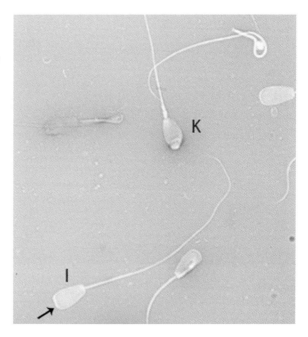

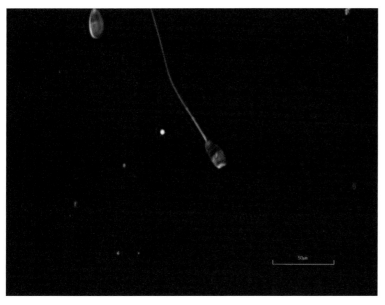

Figure 6.2 Indented acrosome defect with arrowing pointing toward indentation (eosin–nigrosin, bull, 1000×).

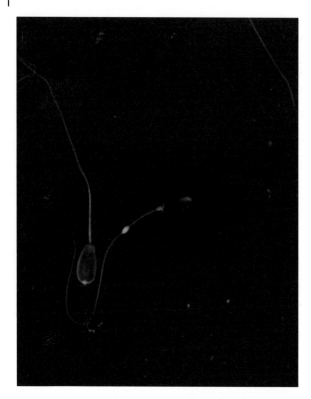

Figure 6.3 Knobbed acrosome defect as noted by arrow, arrowhead points toward distal droplet on adjacent sperm (eosin–nigrosin, bull, 1000×).

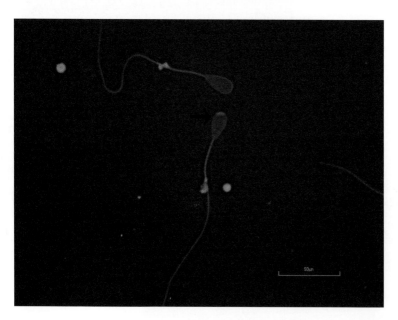

Figure 6.4 Knobbed acrosome as noted by arrow (eosin–nigrosin, bull, 1000×).

Figure 6.5 Indented acrosome as noted by arrow (eosin–nigrosin, bull, 1000×).

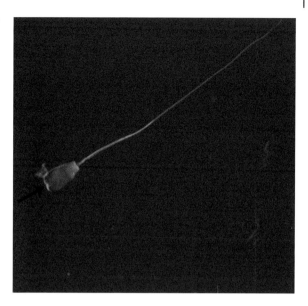

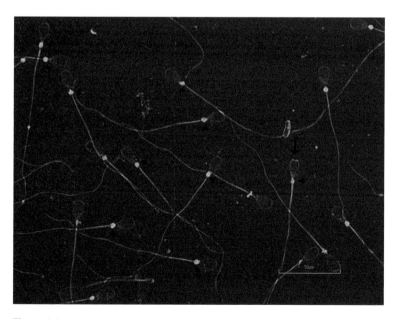

Figure 6.6 Indented acrosome as noted by arrow, arrow heads indicates proximal droplets (eosin–nigrosin, dog, 1000×).

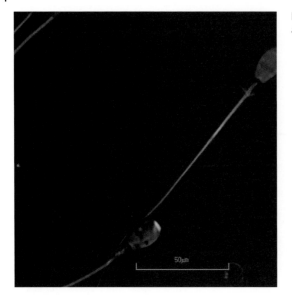

Figure 6.7 Knobbed acrosome defect in a ram as indicated by arrows (eosin–nigrosin, 1000×).

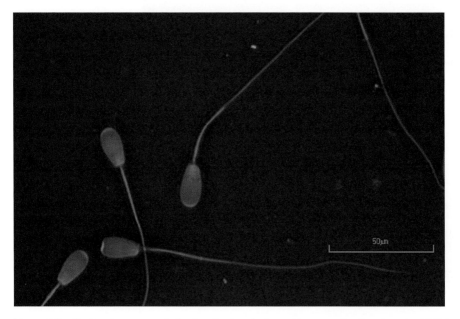

Figure 6.8 Knobbed acrosome defect in a buck as indicated by arrow (eosin–nigrosin, 1000×).

Figure 6.9 Knobbed acrosome defect in a buck as indicated by arrow (eosin–nigrosin, 1000×).

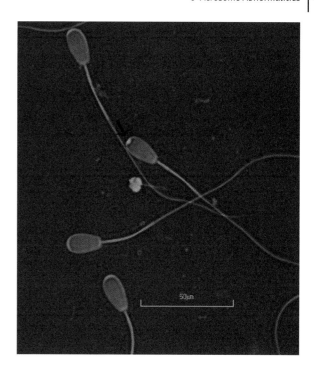

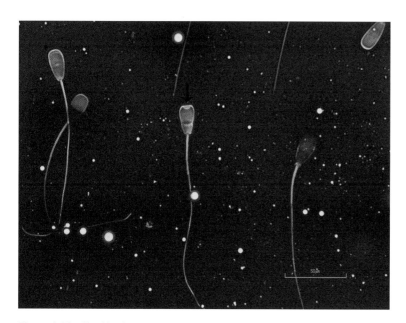

Figure 6.10 Knobbed acrosome defect in a bull in the indented form (eosin–nigrosin, 1000×).

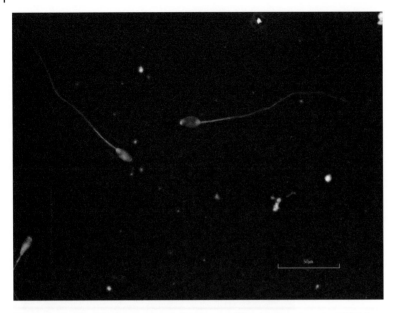

Figure 6.11 Knobbed acrosome defect in a stallion (eosin–nigrosin, 1000×).

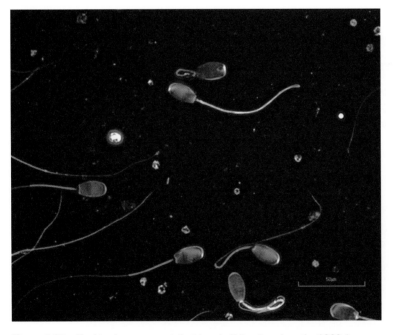

Figure 6.12 Knobbed acrosome defect in a bull (eosin–nigrosin, 1000×).

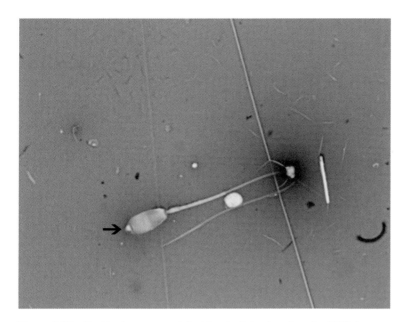

Figure 6.13 Knobbed acrosome defect in a bull (eosin–nigrosin, 1000×).

7

Normal Detached Heads and Free Abnormal Heads

Ejaculates with high percentages of detached heads are associated with reduced fertility and in some cases sterility [89]. Evaluators must try to determine the reason for detached heads as there are several reasons for occurrence. *Detached nonviable but otherwise normal heads* are likely due to senescence of normal sperm during storage in the epididymis or ductus deferens, whereas *detached abnormal heads* may be due to abnormal formation of the basal plate and/or implantation fossa. Alternatively, *detached heads that are "alive"* (vital staining) implies recent head separation; in these cases, the head to midpiece attachment is weak due to abnormal formation of the basal plate or implantation fossa, rather than senescence [29].

The basal plate, a thickening of a part of the caudally directed nuclear envelope that connects the base of the sperm nucleus to the capitulum of the sperm midpiece, forms during the last step of spermiogenesis, just before spermiation [46]. Abnormal chromatin condensation and abnormal morphogenesis of the distal nucleus and basal plate were associated with elevated levels of a variety of sperm abnormalities [90]. Thus, abnormal basal plate formation and an elevation in proportion of detached heads may often be associated with disturbances of spermatogenesis caused by a variety of factors including stress, abnormal testicular thermoregulation, or toxicity [29].

In contrast, **large numbers of detached heads** can occur in association with a **high percentage of non-viable sperm** following abnormal accumulation of senescent sperm in the cauda epididymis and ampullae [91]. In normal males, peristalsis continually moves sperm from the cauda epididymis into the urethra, ensuring a reserve of fresh sperm for ejaculation [92]. Failure of this transport mechanism is associated with sperm accumulation and eventual senescence of stored sperm. Affected bulls are referred to as "sperm accumulators" or "rusty load bulls," whereas stallions are labeled as "sperm accumulators" or suffering from "spermostasis." Characteristics indicative of sperm accumulation are (i) ability to collect large volumes of concentrated semen at one time by electroejaculation or artificial vagina; (ii) normal sperm morphology other than detached heads; (iii) majority sperm nonviable; and (iv) reduction in number of detached heads in subsequent ejaculations [29]. An incidence of 1.1% was reported in Canadian bulls [3]. Comparably, approximately 30–40% of stallions suffer from sperm accumulation or spermiostasis.

Sperm accumulation is due to problems in sperm transport rather than spermatogenesis, and detached sperm heads are due to senescence [31]. With repeated ejaculations, senescent sperm are removed and replaced by a new population. In normal sexually inactive males, excess daily sperm production is voided in the urine [92, 93]. It is postulated that bulls and stallions suffering from sperm accumulation have an irregularity in sperm transport that results in accumulation of spermatozoa in epididymis and ampullae [3, 94].

Sperm Morphology of Domestic Animals, First Edition. J.H. Koziol and C.L. Armstrong.
© 2022 John Wiley & Sons, Inc. Published 2022 by John Wiley & Sons, Inc.
Companion website: www.wiley.com/go/koziol/sperm

Sperm accumulation is not associated with age, and a predilection of re-occurrence of accumulation appears to be permanent [91]. Recurrence within one month of sexual rest is common following resolution of sperm accumulation by frequent electroejaculation [91].

In breeding trials, bulls with a history of sperm accumulation had lower pregnancy rates (50–56%) than control bulls (72–77%). Affected bulls should be used with caution. Following periods of rest, three to five ejaculates may be necessary for the bull to clear accumulated senescent sperm, decreasing the opportunities for some females to become pregnant [91].

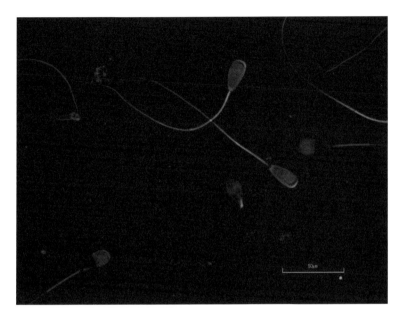

Figure 7.1   Abnormal detached head in a bull (eosin–nigrosin, 1000×).

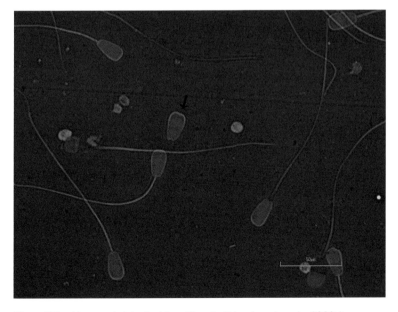

Figure 7.2   Abnormal detached head in a bull (eosin–nigrosin, 1000×).

Similarly in stallions, an average of 8.9 collections over an average of seven days was required to clean out affected stallions to the point where semen values returned to acceptable levels. Management of stallions with spermiostasis requires frequent collections or live breedings to maintain acceptable semen parameters [94]. Love et al. estimated that a 1% increase in the percentage of detached heads resulted in a 2.6% reduction in per cycle pregnancy rates in stallions [32].

In dogs, the presence of increased numbers of detached heads has been associated with testicular degeneration [95, 96].

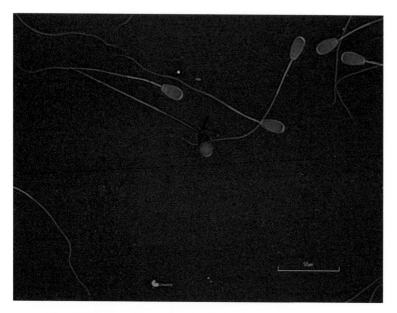

Figure 7.3 Detaching head in a buck (eosin–nigrosin, 1000×).

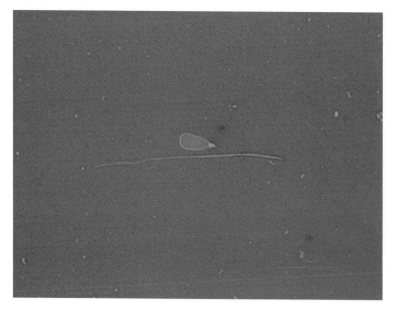

Figure 7.4 Detached abnormal head with corresponding tail in a bull (eosin–nigrosin, 1000×).

8

Decapitated Sperm Defect

The decapitated sperm defect is a rare hereditary defect resulting in infertility. The condition has been reported in Guernsey and Hereford bulls, and the majority of sperm heads in the ejaculate are detached [97–100]. Majority of the loose tails in the ejaculate are bent around a cytoplasmic droplet and often both bent and straight tails are vigorously motile. The cause of the head separation is a defective basal plate and implantation fossa. The sperm head and tail are weakly connected by the cytoplasmic membrane and separation occurs when the tail becomes motile. The three main characteristics associated with the decapitated sperm defect are (i) 80–100% of sperm are affected; (ii) a high percentage of headless tails with progressive motility are present; and (iii) proximal cytoplasmic droplets with or without the tail bending around the droplet are often associated with the motile detached tails. Retained droplets associated with detached tails may be confused with microcephalic heads [101]. In the initial case series describing the defect in sterile Guernsey bulls, there was little doubt that the condition is inherited, likely recessive [100].

Sperm Morphology of Domestic Animals, First Edition. J.H. Koziol and C.L. Armstrong.
© 2022 John Wiley & Sons, Inc. Published 2022 by John Wiley & Sons, Inc.
Companion website: www.wiley.com/go/koziol/sperm

Section II

Midpiece Abnormalities

9

Proximal Cytoplasmic Droplet

Proximal droplets are spherical condensations of cytoplasm 2–3 μm in diameter that surround the neck and proximal midpiece of sperm [102]. These masses of residual cytoplasm move down the principal piece during spermiogenesis and epididymal storage and detach when the sperm are exposed to seminal plasma during ejaculation [62, 102]. A high percentage of sperm with proximal cytoplasmic droplets in an ejaculate is associated with abnormal epididymal function and sperm maturation, or abnormal spermiogenesis, and is often associated with a variety of other sperm defects [3, 103, 104].

Peripubertal males often have a high percentage of sperm with proximal droplets in the ejaculate [3, 104–108]. As males mature, the number of proximal droplets in the spermiogram should decrease [3]. In mature males, the appearance of high numbers of proximal droplets is associated with a degenerative process of the seminiferous epithelium [3]. There is a general consensus that a high percentage of sperm with proximal droplets in a semen sample negatively impacts fertility [103, 109–112]. In a report utilizing sperm with >30% proximal droplets for *in vitro fertilization*, zona binding and cleavage rates were poor [62, 113]. Similar results were found in dogs with impaired capacitation ability being noted by sperm with proximal droplets [114].

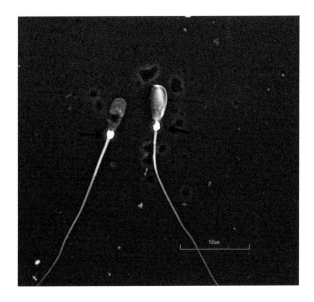

Figure 9.1 Proximal cytoplasmic droplets in a bull – spherical cytoplasmic droplet surrounds the neck and proximal midpiece of sperm. Also, note the vacuole defect in the sperm head on the left and the diadem defect in the sperm on the right (eosin–nigrosin, 1000×).

Sperm Morphology of Domestic Animals, First Edition. J.H. Koziol and C.L. Armstrong.
© 2022 John Wiley & Sons, Inc. Published 2022 by John Wiley & Sons, Inc.
Companion website: www.wiley.com/go/koziol/sperm

Proximal droplets accounted for a significant variation in fertility among stallions with a negative correlation between the percentage of proximal cytoplasmic droplets and per cycle pregnancy rates [33].

Prognosis for recovery depends on the age of the male and other defects noted in ejaculate. Investigation into the root of the sperm abnormalities will allow for a practical prognosis to be reached by the veterinarian.

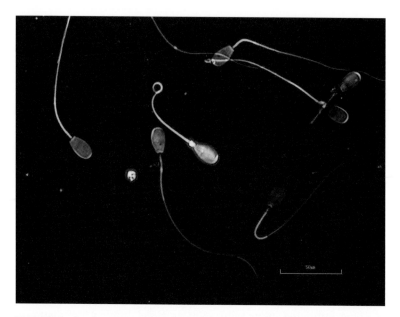

Figure 9.2 Proximal cytoplasmic droplet in a ram as noted by arrow (eosin–nigrosin, 1000×).

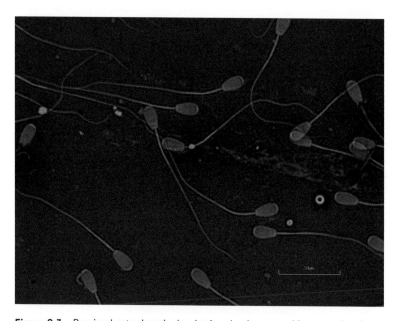

Figure 9.3 Proximal cytoplasmic droplet in a buck as noted by arrow (eosin–nigrosin, 1000×).

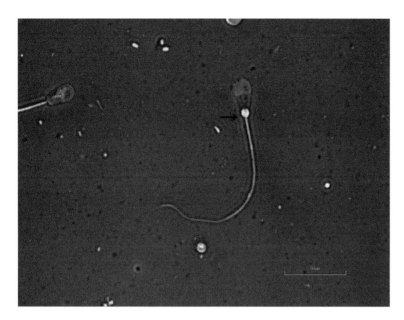

Figure 9.4 Proximal cytoplasmic droplet in a bull (eosin–, 1000×).

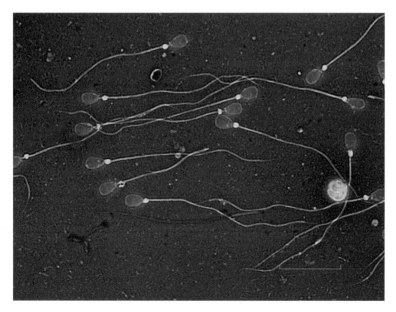

Figure 9.5 Sperm with proximal cytoplasmic droplets in a dog suffering from testicular degeneration (eosin–nigrosin, 1000×).

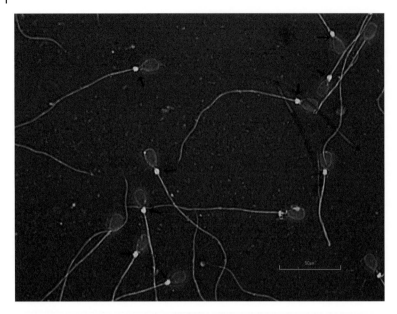

Figure 9.6 Proximal cytoplasmic droplets in a dog as indicated by arrows (eosin–nigrosin, 1000×).

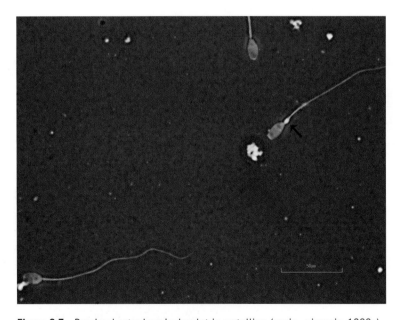

Figure 9.7 Proximal cytoplasmic droplet in a stallion (eosin–nigrosin, 1000×).

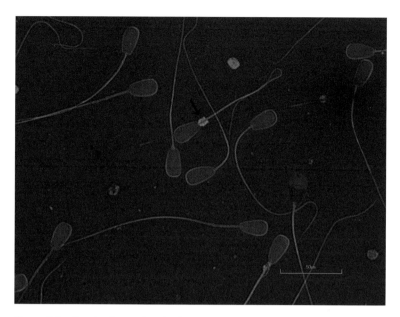

Figure 9.8   Proximal cytoplasmic droplet in a bull (eosin–nigrosin, 1000×).

Figure 9.9   Proximal cytoplasmic droplet in a bull (eosin–nigrosin, 1000×).

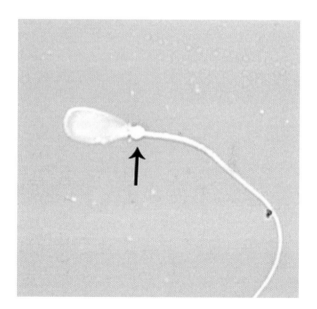

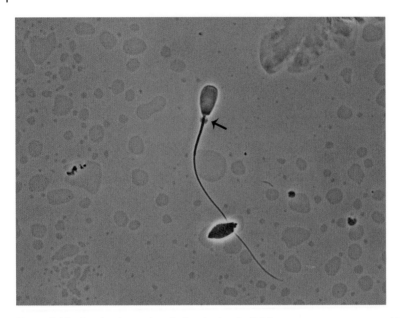

Figure 9.10 Proximal cytoplasmic droplet in a bull (phase contrast wet mount, 1000×).

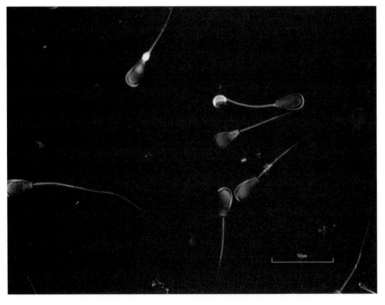

Figure 9.11 Proximal cytoplasmic droplet in a bull as noted by arrow also pyriform-shaped head (eosin–nigrosin, 1000×).

Figure 9.12 Proximal cytoplasmic droplet in a bull (eosin–nigrosin, 1000×).

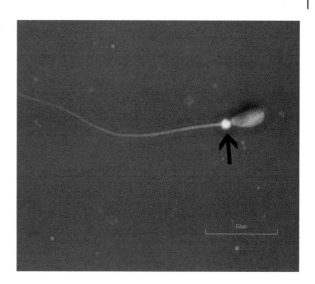

10

Pseudodroplet

The pseudodroplet defect is characterized by a local thickening somewhere along the midpiece and is sometimes referred to as a swollen midpiece. The defective area is often similar in size or smaller than a cytoplasmic droplet and often elongated rather than spherical. In contrast to typical locations of proximal and distal cytoplasmic droplets, pseudodroplets are commonly present in the center of the midpiece, an area where true cytoplasmic droplets are uncommon [115]. The defect may be associated with a bend or fracture of the mitochondrial sheath at the same site. In eosin-nigrosin- or India-ink-stained semen smears, the thickened area may appear to have vacuoles and granules within it. With electron microscopy, affected areas had an accumulation of dense granules on the surface of the mitochondrial helix, surrounded by one or more layers of mitochondria of normal appearance. The origin of the dense granules is unknown [115].

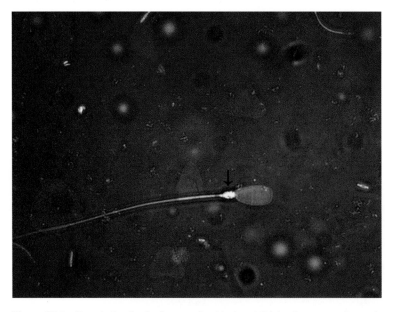

Figure 10.1 Pseudodroplet is characterized by local thickening somewhere along the midpiece. The pseudodroplet typically has vacuoles or a granular appearance (eosin–nigrosin, 1000×).

Sperm Morphology of Domestic Animals, First Edition. J.H. Koziol and C.L. Armstrong.
© 2022 John Wiley & Sons, Inc. Published 2022 by John Wiley & Sons, Inc.
Companion website: www.wiley.com/go/koziol/sperm

11

Mitochondrial Sheath Defects

The mitochondrial sheath is formed during the final steps of spermiogenesis [3]. Mitochondria are laid down first in the sperm neck region and then progress distally toward the annulus at the midpiece-principal piece junction.

A common mitochondria sheath abnormality involves missing mitochondria, resulting in gaps in the midpiece. Large gaps within the mitochondrial sheath result in weakness and a predisposition to fracturing. It is common to find an occasional sperm with one or two small gaps in the midpiece in ejaculates of otherwise fertile males. Rarely, a male may produce a high percentage of affected sperm [29]. Gossypol has been associated with midpiece defects in the bull with aplastic, fragile, or asymmetric midpiece abnormalities noted [116].

Swollen supernumerary mitochondria result in a ragged enlarged area on the midpiece, which may resemble a pseudodroplet defect. With brightfield microscopy, it can be difficult to distinguish among mitochondrial disruptions, proximal cytoplasmic droplets, and pseudodroplets; however, they are more readily identifiable with phase contrast [29].

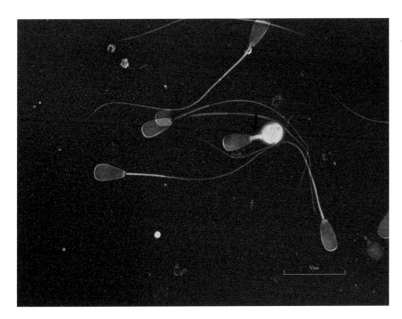

Figure 11.1 Abnormal midpiece that culminates in a terminally coiled tail (eosin–nigrosin, bull 1000×).

Sperm Morphology of Domestic Animals, First Edition. J.H. Koziol and C.L. Armstrong.
© 2022 John Wiley & Sons, Inc. Published 2022 by John Wiley & Sons, Inc.
Companion website: www.wiley.com/go/koziol/sperm

Microtubular mass defects of sperm have been described in stallions, which resemble pseudodroplets, corkscrew defect, and cytoplasmic droplets [117]. Seven Standardbred stallions were described with this defect with 4/7 stallions being considered subfertile [118]. A similar midpiece defect termed "knobs" or mitochondrial accumulations were observed in a stallion and was characterized as having a thick, rough midpiece that was overlaid with several layers of mitochondria

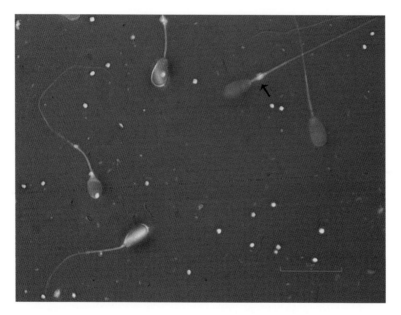

Figure 11.2 Abnormal midpiece in a bull (eosin–nigrosin, 1000×).

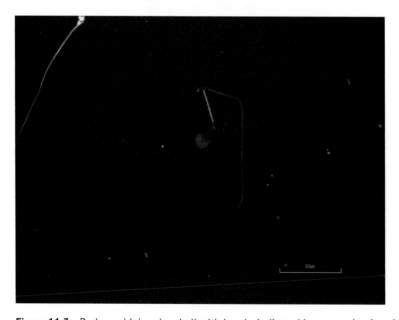

Figure 11.3 Broken midpiece in a bull with breaks indicated by arrows (eosin–nigrosin, 1000×).

of different shapes, sizes, and orientation to the long axis of the sperm [119, 120]. Similar reports of mitochondrial sheath defects have also been described in the boar [121].

Mitochondrial helix segmental aplasia commonly occurs at low levels in ejaculates of some males; however, little is known about the cause. It is uncommon to see this defect representing a high percentage of abnormalities in a spermiogram.

Figure 11.4 Broken midpiece at the site of attachment in a bull (eosin–nigrosin, 1000×).

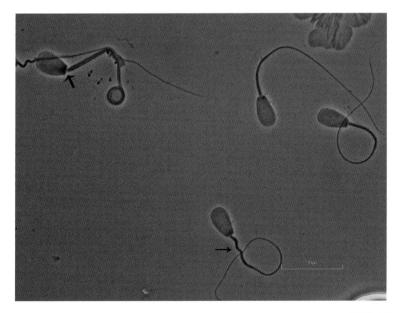

Figure 11.5 Abnormal midpieces in a bull (phase contrast wet mount, 1000×). The large arrow spermatozoa is also a macrocephalic sperm. The smaller arrow has a diadem defect.

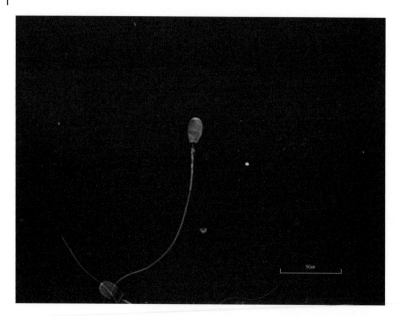

Figure 11.6 Abnormal midpiece in a ram, note the roughened mitochondrial sheath (eosin–nigrosin, 1000×).

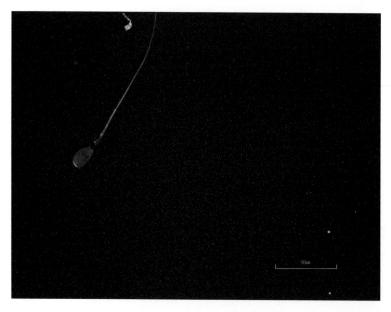

Figure 11.7 Abnormal midpiece with mitochondrial sheath defect in a ram (eosin–nigrosin, 1000×).

Figure 11.8 Mitochondrial sheath defects as indicated by arrows in a ram (eosin–nigrosin, 1000×).

Figure 11.9 Mitochondrial sheath defect in a bull (eosin–nigrosin, 1000×).

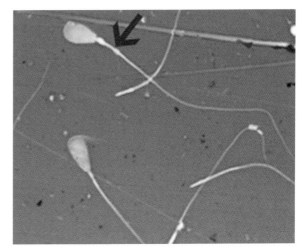

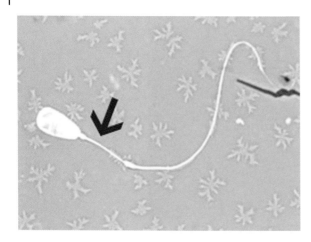

Figure 11.10 Mitochondrial sheath defect in a bull (eosin–nigrosin, 1000×).

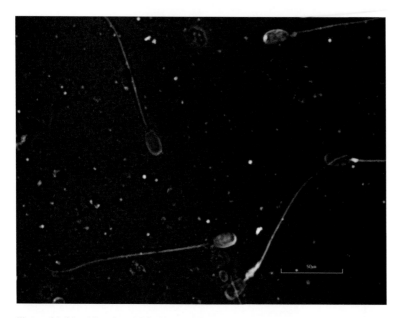

Figure 11.11 Mitochondrial sheath defect in a buck (eosin–nigrosin, 1000×).

12

Corkscrew Sperm Defect

The corkscrew sperm defect was first described by Blom in 2 sterile Red Danish bulls [122] and later in 60 other Danish bulls of various breeds and a sterile stallion [119, 123]. This defect is characterized by an irregular distribution of mitochondria along the length of the mitochondrial sheath and are often described as "roughed up" midpieces. Electron microscopy displays sporadic mitochondrial swelling, degeneration, and aplasia along the entire length of the sheath. Affected sperm may also have a proximal droplet. Despite the weakened condition of the midpiece, folding or coiling of the tail does not occur in the corkscrew defect. The classic appearance of the defect is present only in nonviable cells, because in motile cells, the midpiece would fracture at initiation of flagellar activity [29].

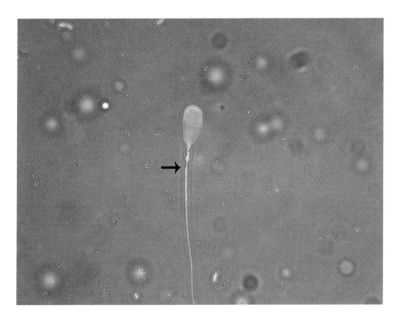

Figure 12.1 Corkscrew sperm in a bull as indicated by arrow (eosin–nigrosin, 1000×).

Sperm Morphology of Domestic Animals, First Edition. J.H. Koziol and C.L. Armstrong.
© 2022 John Wiley & Sons, Inc. Published 2022 by John Wiley & Sons, Inc.
Companion website: www.wiley.com/go/koziol/sperm

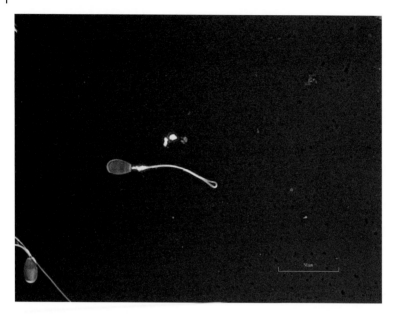

Figure 12.2   Corkscrew sperm defect in a ram (eosin–nigrosin, 1000×).

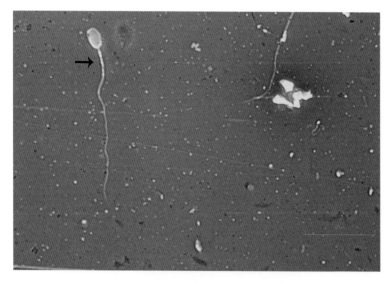

Figure 12.3   Corkscrew defect in buck (eosin–nigrosin, 1000×).

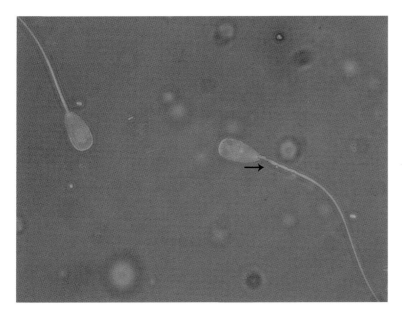

Figure 12.4 Corkscrew defect in bull (eosin–nigrosin, 1000×).

13

Dag Defect

The "Dag" defect is characterized by multiple fractures of the axonemal fibers in the sperm mid-piece, causing mitochondrial sheath disruption. The defect was named after the Jersey bull "Dag" that, along with his full brother, produced semen with high percentages of affected sperm [124]. The Dag defect has been demonstrated as heritable in Jerseys, Swedish Red and White, [125] Hereford, [126] and Galloway bulls. A similar and perhaps indistinguishable defect may occasionally be present in a few sperm in an ejaculate of a variety of species due to a disturbance of spermatogenesis [29].

When nearly all sperm are affected, the Dag defect is likely of genetic origin and inherited as a simple recessive trait [127]. Key features of the Dag defect are a swollen disrupted mitochondrial helix, multiple fractures of the midpiece, often with coiling of the principal piece around a cytoplasmic droplet or folding and coiling of the midpiece with the axis of the main fold in the distal half of the midpiece [3]. Sperm leaving the testis and entering the caput epididymis have an

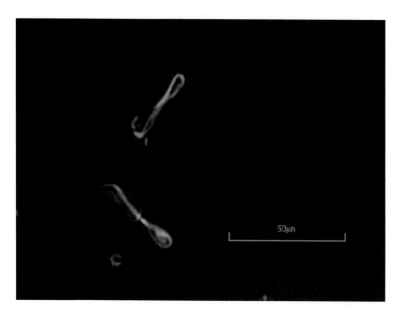

Figure 13.1 Dag defect as indicated by arrow with distal midpiece reflex below in a bull (eosin–nigrosin, 1000×).

Sperm Morphology of Domestic Animals, First Edition. J.H. Koziol and C.L. Armstrong.
© 2022 John Wiley & Sons, Inc. Published 2022 by John Wiley & Sons, Inc.
Companion website: www.wiley.com/go/koziol/sperm

irregular but intact mitochondrial sheath. As sperm traverse the epididymis, progressive changes in morphology can be first detected in the distal part of the caput epididymis, due to a structural weakness in the mitochondrial sheath and cementing material of the midpiece [3, 128].

Affected Jersey bulls appeared to have decreased severity of midpiece disruption compared to other breeds and may have a separatse mode of inheritance [3]. Severely affected bulls are sterile and affected bulls do not recover [29].

A similar defect to that commonly seen in the bull was described in a Standardbred stallion that suffered subfertility by Hellander et al. [129] as well as an English Bulldog by Rota et al. [130].

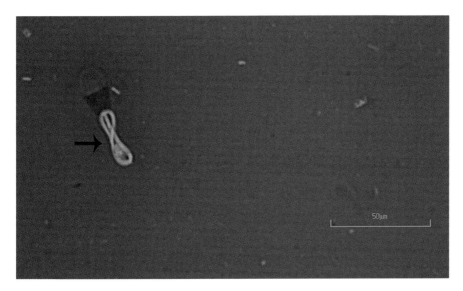

Figure 13.2   Dag defect in a bull (eosin–nigrosin, 1000×).

Figure 13.3   Dag defect in a bull (eosin–nigrosin, 1000×).

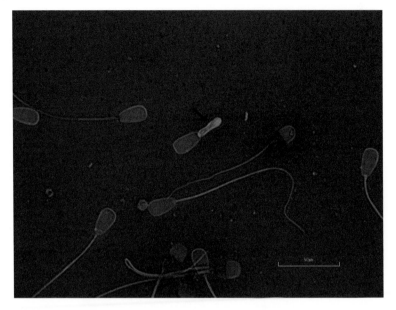

Figure 13.4   Dag defect in a bull (eosin–nigrosin, 1000×).

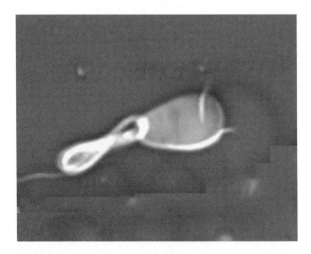

Figure 13.5   Dag defect in a bull (eosin–nigrosin, 1000×).

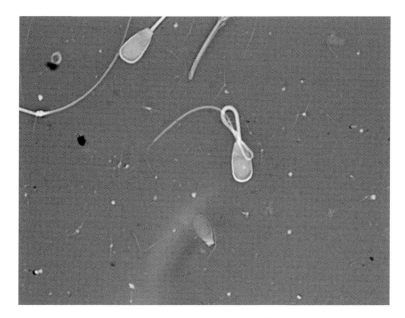

Figure 13.6 Dag defect in a bull (eosin–nigrosin, 1000×).

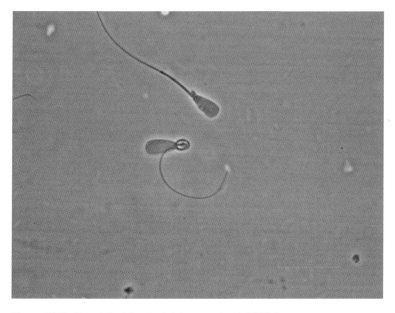

Figure 13.7 Dag defect in a bull (phase contrast, 1000×).

14

Distal Midpiece Reflex

The distal midpiece reflex (DMR) is a common sperm midpiece abnormality in most species and the most prevalent midpiece abnormality present in the semen of bulls. The typical appearance under light microscopy is of a bend in the distal midpiece in the shape of the letter "J" or a shepherds crook. A cytoplasmic droplet is almost always trapped in the bend of the midpiece and the principal piece extends beyond the sperm head. During motility assessment, affected sperm are often swimming backwards and/or in tight circles. Variations in bending result in various manifestations of the same defect, including an L-shaped bend and a second bend in the opposite direction more proximally in the midpiece [29]. A significant difference between a DMR and the Dag defect is that the DMR has a smooth and complete mitochondrial sheath as opposed to the roughened and fractured mitochondrial sheath of the Dag defect [3].

DMR defects are due to an abnormal environment in the cauda epididymis, specifically, the distal third of the cauda epididymis [3]. High concentrations of testosterone are necessary for normal epididymal function [131]. A stress-induced rise in cortisol decreases secretion of luteinizing hormone (LH) and testosterone and interferes with normal epididymal function [6, 8]. The resultant abnormal epididymal function leads to an increase in DMRs secondary to low testosterone, which is secondary to a variety of adverse conditions including estradiol treatment, [132, 133] low thyroid activity, [134] ephemeral fever, [135] scrotal insulation, [136] or simply a cortisol release associated with any stressful event of sufficient duration (e.g. snowstorm) [3]. Following a stressful event, the percentage of DMRs increases rapidly within one week after the onset of the stress with numbers peaking by the second week [3, 137]. The initial rise of DMRs occur in the population already in the epididymis during the event. The population of spermatozoa cells was about to enter the epididymis during the stressful event. If the stress is mild and short-lived, percentages of DMRs in the spermiogram will soon regress. However, if the stress was more severe and prolonged, other types of defects will follow the DMRs and recovery of the spermiogram will take at least six to eight weeks.

DMRs are a seasonal problem for many bulls in northern regions [138]. This may be related to photoperiodic changes in testosterone and/or stress secondary to winter weather conditions. Barth et al. reported that testosterone concentrations are lower in fall and winter, but increase as spring approaches and reach peak concentrations in June [139]. Barth reports that it is not uncommon for some bulls to produce semen with 60–80% DMRs due to effects of low testosterone and/or winter stress, with Angus bulls appearing more predisposed to the defect and more apt to maintain the abnormality into late spring [29].

Sperm Morphology of Domestic Animals, First Edition. J.H. Koziol and C.L. Armstrong.
© 2022 John Wiley & Sons, Inc. Published 2022 by John Wiley & Sons, Inc.
Companion website: www.wiley.com/go/koziol/sperm

Since DMRs result from disruptions of sperm located within the cauda epididymis, where most sperm normally still have a cytoplasmic droplet in the distal position of the midpiece, droplets are almost always entrapped in the fold of the midpiece [137]. Sperm suffering from cold or hypotonic shock should not be mistaken for DMR. The occurrence of a large number of bent tails **without** a retained cytoplasmic droplet on routine motility and morphology evaluations should alert the evaluator that an iatrogenic process may be in play.

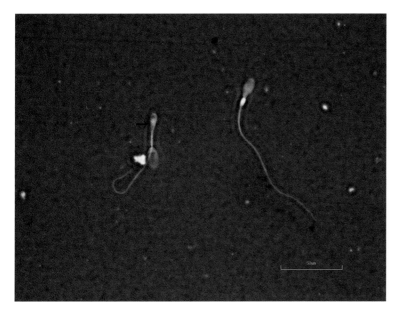

Figure 14.1 Distal midpiece reflex (DMR) in a stallion. Note proximal droplet to the right (eosin–nigrosin, 1000×).

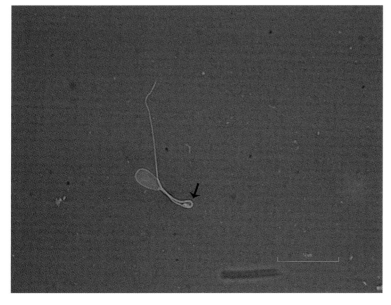

Figure 14.2 DMR in a bull. Note midpiece reflexed around a droplet, which is a hallmark of a DMR (eosin–nigrosin, 1000×).

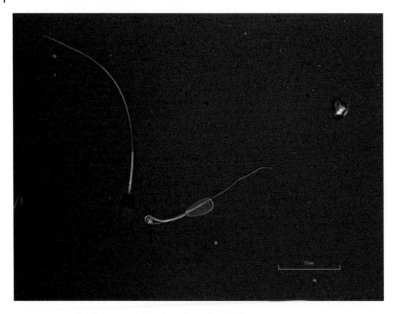

Figure 14.3 DMR in a bull (eosin–nigrosin, 1000×).

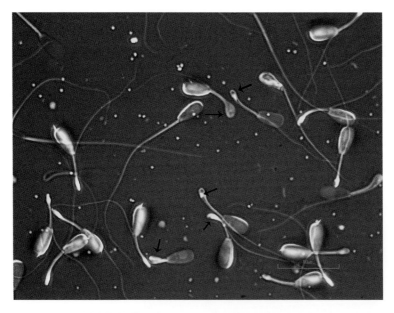

Figure 14.4 Multiple DMR defects in a bull suffering from environmental stress (eosin–nigrosin, 1000×).

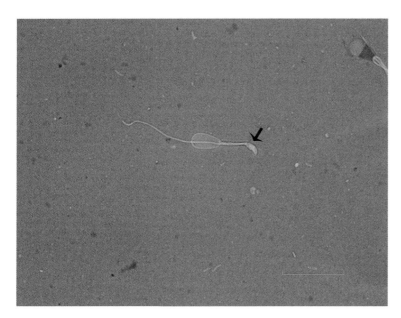

Figure 14.5   DMR in a bull (eosin–nigrosin, 1000×).

Figure 14.6   DMR in a bull
(eosin–nigrosin, 1000×).

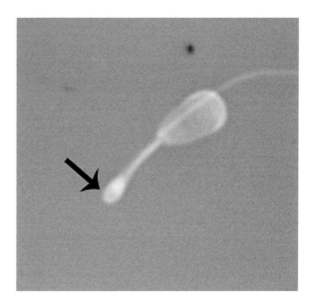

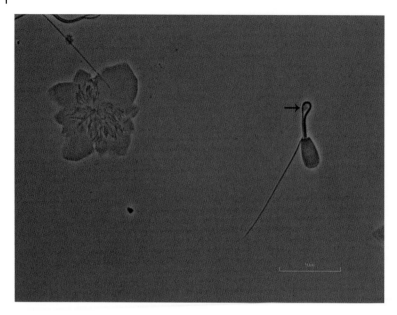

Figure 14.7 DMR in a bull, note the trapped droplet (phase contrast wet mount, 1000×).

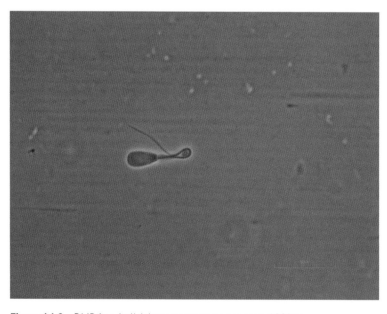

Figure 14.8 DMR in a bull (phase contrast wet mount, 1000×).

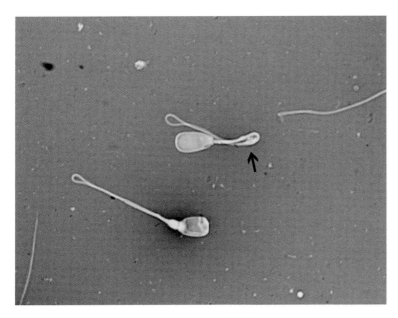

Figure 14.9 DMR in a bull (eosin–nigrosin, 1000×).

15

Bowed Midpieces

Bowed midpieces are usually artifacts that result when sperm are immobilized during a flagellar contraction as sperm dries during staining. In rare cases, large percentages of sperm may be affected and a stiff circling movement of sperm may be apparent on a wet mount. Sperm motion must be assessed (wet mount) to determine whether bowed midpieces observed in high numbers on stained smears are an artifact or due to a structural abnormality [29].

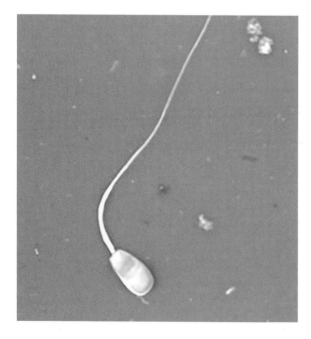

Figure 15.1 Bowed midpiece (eosin–nigrosin, 1000×).

Sperm Morphology of Domestic Animals, First Edition. J.H. Koziol and C.L. Armstrong.
© 2022 John Wiley & Sons, Inc. Published 2022 by John Wiley & Sons, Inc.
Companion website: www.wiley.com/go/koziol/sperm

16

Distal Droplets

Distal droplets are characterized as cytoplasmic droplets located just proximal to the annulus. The implication of distal droplets on fertility appears to differ based on species and research studies. Some studies do not differentiate between proximal vs. distal cytoplasmic droplets, which cause some difficulty in interrupting the implications of distal droplets on fertility in those studies. Distal droplets may be associated with immaturity in most species and may indicate a disturbance in the caudal epididymis in certain animals and is often in the presence of other sperm abnormalities that corroborate the diagnosis.

In bulls, approximately 35% of sperm shed distal droplets while still in the caudal epididymis with the remaining sperm shedding the droplet upon mixing with fluid from the accessory sex glands [3]. Distal droplets have not been consistently implicated in substantial impairment of bull fertility in natural service settings, even when present at high percentages in the ejaculate. Consequently, sperm with distal droplets and no other discernible abnormality should not be regarded as abnormal in bulls intended for natural service [3, 140].

In a Swedish study, bulls that produced high percentages of sperm with distal droplets in their ejaculates had artificial insemination (AI) non-return rates similar to cohorts with few distal droplets in their ejaculate [141]. Another study using Zebu bulls suggested distal droplets were inversely related to age, with immature bulls displaying higher percentages, and mature and older bulls having a lower prevalence [142].

Normally, when semen is mixed with seminal fluid at ejaculation, sperm become motile and droplets are shed from the distal midpiece due to the increased fluidity of the sperm membrane over the droplet region. At the distal midpiece region, the membrane stretches to form a stalk and then ruptures, releasing a non-membrane-bound remnant of spermatid cytoplasm. Some hypothesize retention of the distal cytoplasmic droplets may be due to a lack of a hemolytic factor, specifically, phospholipid binding protein, in seminal fluid [140, 143, 144]. Much background debris in semen smears is from shed or disintegrating droplets [29].

One theory concerning the cause of distal droplets in ejaculate is insufficient seminal fluid release at ejaculation [29]. Interestingly, when semen with a high number of distal droplets was incubated for 15–30 min or the sample was gently agitated, retained distal droplets were released from the sperm [3].

In boars, it is recommended that semen containing >15% proximal and distal cytoplasmic droplets should not be used for artificial insemination [145]. Waberski et al. found that proximal and distal droplets were negatively correlated with pregnancy rates and litter size in dogs. Increase in distal droplets have been noted in males intensively mated [146].

Sperm Morphology of Domestic Animals, First Edition. J.H. Koziol and C.L. Armstrong.
© 2022 John Wiley & Sons, Inc. Published 2022 by John Wiley & Sons, Inc.
Companion website: www.wiley.com/go/koziol/sperm

Retained cytoplasmic droplets may be responsible for loss of membrane fluidity and integrity that compromise the cryotolerance of sperm with this specific defect and may account for differences in fertility rates between frozen semen containing distal droplets versus fresh ejaculates [147].

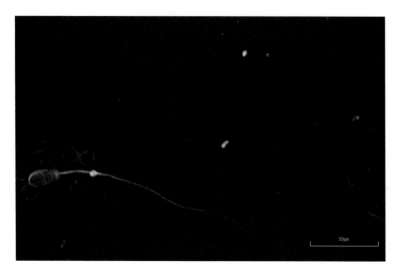

Figure 16.1　Distal droplet in a bull (eosin–nigrosin, 1000×).

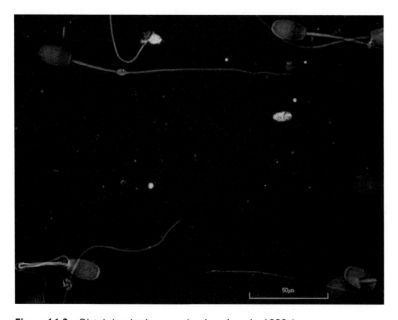

Figure 16.2　Distal droplet in a ram (eosin–nigrosin, 1000×).

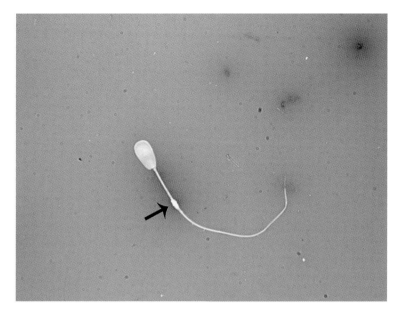

Figure 16.3 Distal droplet in a bull (eosin–nigrosin, 1000×).

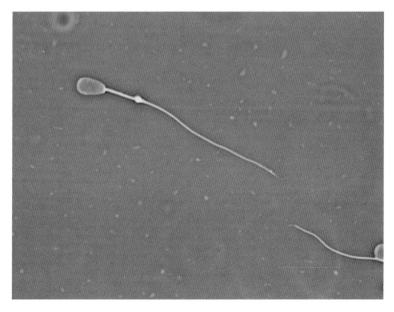

Figure 16.4 Distal droplet in a bull (phase contrast wet mount, 1000×).

17

Abaxial Midpieces

Abaxial tail attachment is not considered abnormal in the stallion, [148] boar, [149] dog, [103] or bull; [3, 150] therefore, these sperm should be considered a normal morphological variation. Combined results of experiments to determine effects of abaxial tails on fertility indicate that sperm with abaxial tails fertilize ova at a normal rate and are not associated with increased embryonic death [150].

Abaxial attachments of the sperm tail to the base of the nucleus occur in low to very high proportions of sperm, either alone or in conjunction with accessory tails. Sperm with abaxial tails usually have an empty implantation fossa. Abaxial tails do not appear to originate as a result of extrinsic factors. Males producing sperm with abaxial tails often do not display other sperm abnormalities in high percentages, and the presence of the abaxial tails remains constant over long intervals [3]. When ejaculates contain sperm with abaxial tails but limited other sperm abnormalities, males generally have grossly normal testes and produce semen with normal sperm concentration and progressive motility that can be successfully cryopreserved [29].

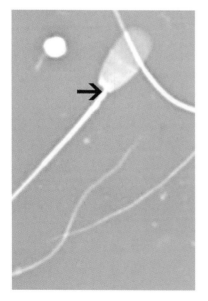

Figure 17.1   Abaxial implantation (arrow) (eosin–nigrosin, bull, 1000×).

Sperm Morphology of Domestic Animals, First Edition. J.H. Koziol and C.L. Armstrong.
© 2022 John Wiley & Sons, Inc. Published 2022 by John Wiley & Sons, Inc.
Companion website: www.wiley.com/go/koziol/sperm

Figure 17.2 Abaxial implantation in a stallion (eosin–nigrosin, 1000×).

Section III

Tail (Principal Piece) Abnormalities

18

Tail Stump Defect

The tail stump defect is an uncommon sperm defect reported in multiple breeds of cattle [151–153]. A similar defect has been reported in other species including the mouse, rabbit, dog, stallion, and man but is usually found in low percentages [3]. Utilizing light microscopy, examination of semen smears may give an initial impression of a *very high* incidence of detached heads; however, closer examination reveals the tail is replaced by a small stump at the base of the sperm head [29]. The stump is often partially obscured by a retained proximal cytoplasmic droplet. Semen produced by affected males usually has a low sperm concentration and poor motility. The tail stump defect is often in association with other defects such as Dag defects, nuclear vacuoles, and pyriform heads. Very few loose tails will be found in spermiograms, since no tail is formed for most affected sperm. The tail stump defect is thought to result from an inherited recessive genetic factor. Affected bulls are usually completely sterile and would not be expected to recover [29].

Figure 18.1 Tail stump defect with stumps obscured by cytoplasmic droplets (eosin–nigrosin, bull, 1000×).

Sperm Morphology of Domestic Animals, First Edition. J.H. Koziol and C.L. Armstrong.
© 2022 John Wiley & Sons, Inc. Published 2022 by John Wiley & Sons, Inc.
Companion website: www.wiley.com/go/koziol/sperm

Figure 18.2 Tail stump defect in a bull (eosin–nigrosin, 1000×).

Figure 18.3 Tail stump defect in a bull, also note pyriform shape and diadem defects (eosin–nigrosin, 1000×).

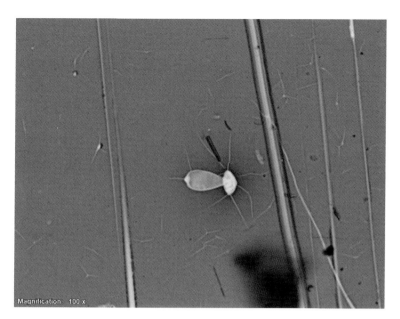

Figure 18.4 Tail stump defect in a bull (eosin–nigrosin, 1000×).

19

Coiled Principal Pieces

The coiled principal piece defect is characterized by tight coiling of the principal piece around a cytoplasmic droplet at various levels along its length [29]. Coiled principal pieces have been reported in most species [29, 32, 154]. Although commonly encountered when evaluating spermiograms, coiled principal pieces seldom exceed 1–5%. Love et al. reported that a 1% increase in the percentage of coiled tails resulted in a 3.9% reduction in per cycle pregnancy rates [32].

A small number of bulls are predisposed to producing the defect, and in these bulls as many as 50% of sperm may be affected following stress or abnormal testicular thermoregulation. Experimental evidence suggests that this defect forms during spermatogenesis; however, it often appears concurrently in spermiograms with defects having an epididymal origin, such as the distal midpiece reflex. There may be a hereditary predisposition for coiled principal pieces, as several bulls have been identified that produce the defect in high percentages. Barth reported this defect in many breeds of bulls; however, Hereford bulls seem to be commonly represented [29]. It may be that certain individuals or breeds are predisposed to developing coiled principal pieces precipitated by low testosterone concentrations due to season, stress, or poor testis thermoregulation [29].

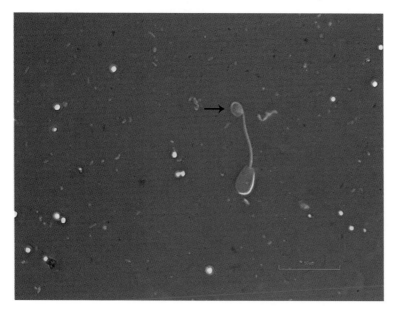

Figure 19.1 Coiled principal piece in a bull (eosin–nigrosin, 1000×).

Sperm Morphology of Domestic Animals, First Edition. J.H. Koziol and C.L. Armstrong.
© 2022 John Wiley & Sons, Inc. Published 2022 by John Wiley & Sons, Inc.
Companion website: www.wiley.com/go/koziol/sperm

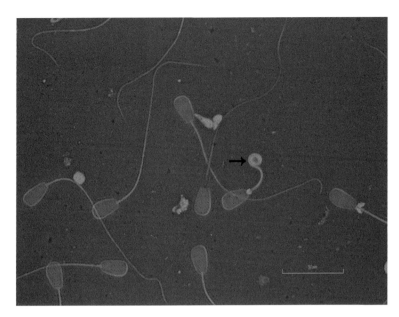

Figure 19.2   Coiled principal piece in a bull (arrow) sperm also has a proximal droplet (eosin–nigrosin, 1000×).

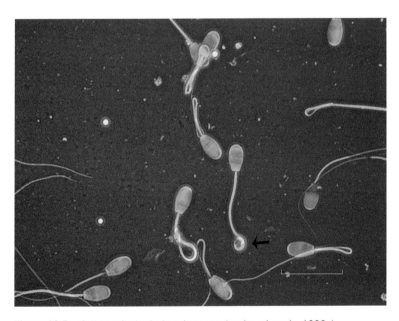

Figure 19.3   Coiled principal piece in a ram (eosin–nigrosin, 1000×).

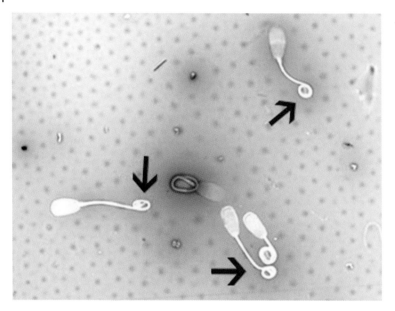

Figure 19.4   Coiled principal piece in a bull (arrows) (eosin–nigrosin, 1000×).

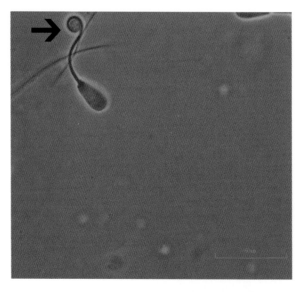

Figure 19.5   Coiled principal piece in a bull (arrow) (phase contrast wet mount, 1000×).

20

Double Forms and Accessory Tails

During spermatogenesis, tail formation begins with migration of the proximal and distal centrioles to the base of the round spermatid nucleus [29]. The implantation fossa forms in advance of the arrival of the pair of migrating centrioles. The proximal centriole arranges itself perpendicular to the distal centriole and becomes the capitulum, whereas the distal centriole gives rise to the tail of the sperm. In normal spermatids, centriole replication is suppressed so that only one flagellum develops. If there is failure of suppression or partial suppression of centriolar division, accessory tails or double tails may occur. The proximal centriole is an important structure in embryonic development, as it forms one of the asters involved in chromosome separation during the first cleavage after fertilization [155–158].

Formation of accessory tails is uncommon and appears to have a heritable basis. When viewed using light microscopy, accessory tails appear as short stumps ending in a filamentous strand. The accessory tail is often obscured by a cytoplasmic droplet, and when accessory tails are present, the incidence of abaxial implantation and double tails is increased [29]. Feulgen-stained slides are preferred to eosin–nigrosin preparations for detecting accessory tails or double implantation fossas [3]. Accessory tails do not appear to occur as the result of extrinsic factors that have an adverse

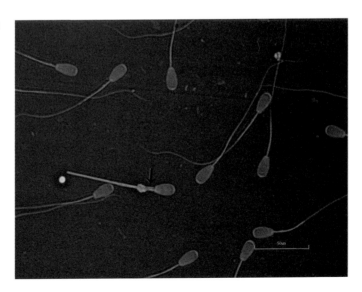

Figure 20.1 Double midpiece in a buck (eosin–nigrosin, 1000×).

Sperm Morphology of Domestic Animals, First Edition. J.H. Koziol and C.L. Armstrong.
© 2022 John Wiley & Sons, Inc. Published 2022 by John Wiley & Sons, Inc.
Companion website: www.wiley.com/go/koziol/sperm

effect on spermatogenesis. It is not known if sperm with accessory tails could fertilize ova; however, if such sperm did enter the ovum, an accessory tail would potentially supply an extra aster, resulting in abnormal chromosome separation at first cleavage [29].

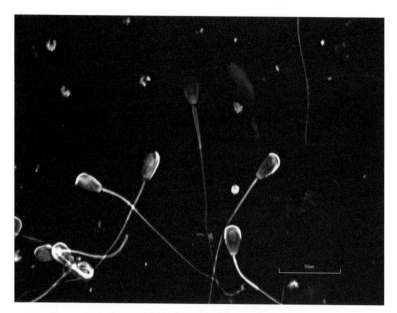

Figure 20.2 Double midpieces in a bull (eosin–nigrosin, 1000×).

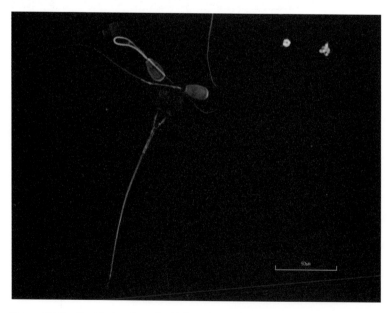

Figure 20.3 Double heads and midpiece that culminate into a single principal piece (eosin–nigrosin, 1000×).

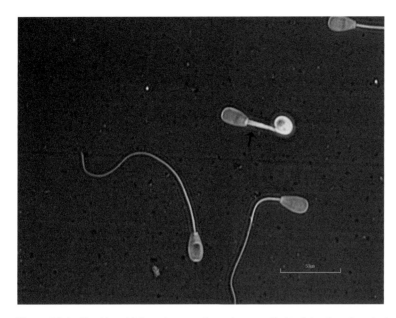

Figure 20.4   Double midpiece that terminate into a coiled tail (eosin–nigrosin, bull, 1000×).

Figure 20.5   Double midpieces in a bull (eosin–nigrosin, 1000×).

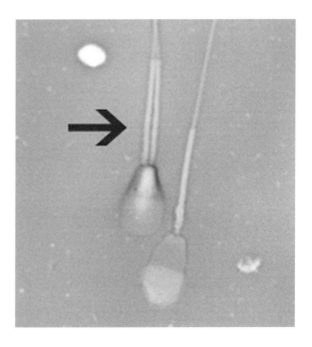

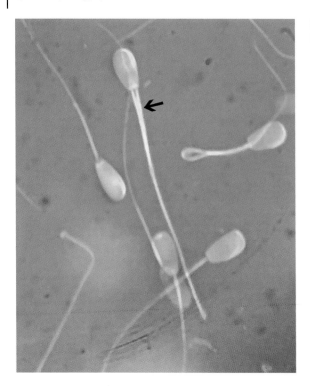

Figure 20.6 Double midpiece in a bull indicated by arrow (eosin–nigrosin, 1000×).

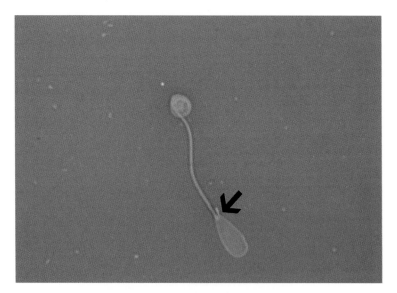

Figure 20.7 Accessory tail indicated by arrow in a bull (eosin–nigrosin, 1000×).

21

Bent Principal Pieces

Sperm with a loop-like bend in the principal piece just distal to the annulus occur in small numbers, often in association with distal midpiece reflexes. Hypotonic shock may cause a similar bend, albeit without a trapped droplet.

Figure 21.1 Bent principal piece in a bull – note the lack of cytoplasmic droplet (eosin–nigrosin, 1000×).

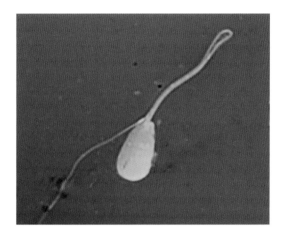

Figure 21.2 Bent principal piece in a bull as indicated by arrowhead compare to the distal midpiece reflex (arrow) that contains a cytoplasmic droplet (eosin–nigrosin, 1000×).

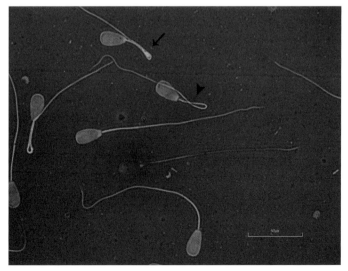

Sperm Morphology of Domestic Animals, First Edition. J.H. Koziol and C.L. Armstrong.
© 2022 John Wiley & Sons, Inc. Published 2022 by John Wiley & Sons, Inc.
Companion website: www.wiley.com/go/koziol/sperm

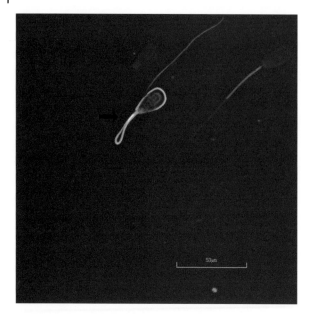

Figure 21.3   Bent principal piece in a bull (eosin–nigrosin, 1000×).

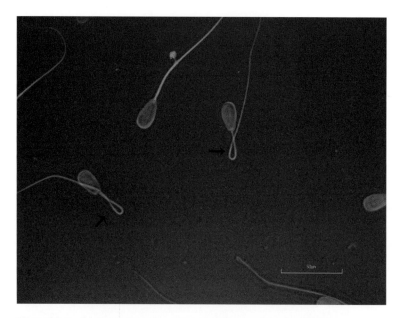

Figure 21.4   Bent principal pieces in a bull (eosin–nigrosin, 1000×).

22

Short Tail Defect

The short tail defect is described in boars and in bulls and is characterized by a markedly reduced or missing principal piece in all sperm, resulting in a total lack of sperm motility [121, 159–161]. Spermiograms from three Nelore bulls affected with this condition have also been described, two of which were half siblings. Along with the short tails, 1–6% proximal droplets were presented along with a small percentage of other defects, with no indications of environmentally based disturbances in spermatogenesis [159]. The defect has been linked to an autosomal recessive gene in boars [161]. The autosomal recessive disease is associated with the insertion of the KPL2 gene in porcine chromosome 16. Inherited short tail syndrome (ISTS) was found to affect 7.6% of Yorkshire boars in one study [162].

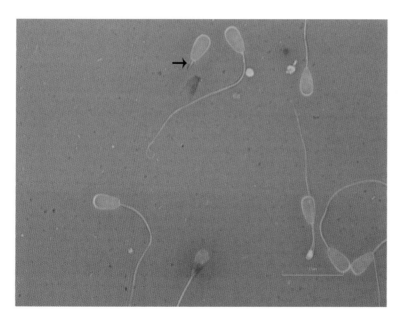

Figure 22.1 Short tail defect in a bull characterized by missing principal piece (eosin–nigrosin, 1000×).

Sperm Morphology of Domestic Animals, First Edition. J.H. Koziol and C.L. Armstrong.
© 2022 John Wiley & Sons, Inc. Published 2022 by John Wiley & Sons, Inc.
Companion website: www.wiley.com/go/koziol/sperm

Figure 22.2 Short tail defect in a bull (eosin-nigrosin, 1000×).

Section IV

Aberrations of Stains and Other Cells in the Ejaculate

23

Aberrations Due to Staining

Hypotonic stains such as eosin–nigrosin and aniline blue may cause distortion of the principal piece and midpiece when slide preparation technique is poor. Cold slides, prolonged drying of smears, chilling semen before staining, and urine contamination may all result in hypotonic shock. Most commonly there will be severe bending or folding of the distal portion of the principal piece. In severe cases, many sperm may resemble distal midpiece reflexes, with the appearance of the J-shaped end at the distal midpiece; however, there will be no trapped droplet material (Figure 23.1).

Hypotonic shock can be minimized by drying smears quickly. Warm slides and blowing air over the smear expedite the drying process. If hypotonic shock occurs despite proper slide preparation, osmolality of the eosin–nigrosin stain should be checked to ensure that the solution has not become excessively hypotonic. If hypotonicity of the morphology stain is discovered, the bottle can be discarded and replaced with a fresh bottle. Alternatively, the osmolality of eosin–nigrosin stain may be increased by the addition of glucose, sodium, or TES-Tris [3, 140].

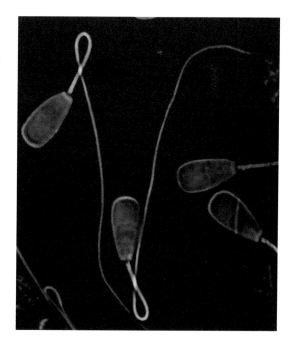

Figure 23.1 When multiple sperm with bent principal pieces and no retained droplet are noted during morphology evaluation concerns of hypo-osmolality of stain or prolonged drying of stain should be considered (eosin–nigrosin, bull, 1000×).

Sperm Morphology of Domestic Animals, First Edition. J.H. Koziol and C.L. Armstrong.
© 2022 John Wiley & Sons, Inc. Published 2022 by John Wiley & Sons, Inc.
Companion website: www.wiley.com/go/koziol/sperm

One might note other aberrations in the stained slide during evaluations such as cracks in the stain or background debris, which could include epithelial cells or other cellular debris, and particles of undissolved stain. Breakdown products of shed cytoplasmic droplets, white blood cells, red blood cells, bacteria, and other microorganisms may also apparent in the background. Cracks in the stain background may occur when an excessive amount of stain is used and the slide is not dried quickly. Cracks often form along the edge of the sperm head extending past its apical margin. Cracks frequently occur in stored slides as the stain dries [29]. If eosin-nigrosin-stained slides are intended to be stored long term, it is a good idea to use mounting glue to fasten a coverslip.

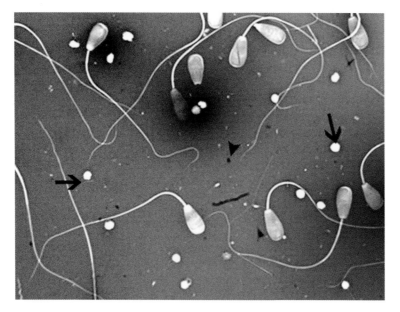

Figure 23.2 Cytoplasmic droplets are shed during ejaculation and can be noted as background debris on stained slides (arrows) (eosin–nigrosin, bull, 1000×).

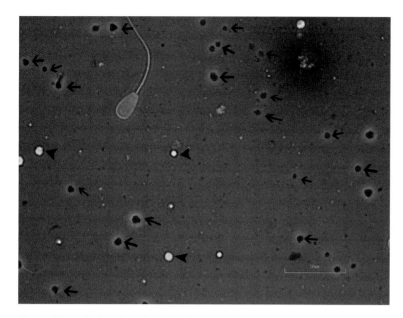

Figure 23.3 Eosin–nigrosin stain often contains small particles of undissolved stain (arrows). Cytoplasmic droplets shed by sperm during emission and ejaculation may be noted, along with disintegrated cytoplasmic droplets as background debris (arrow head) (eosin–nigrosin, bull, 1000×).

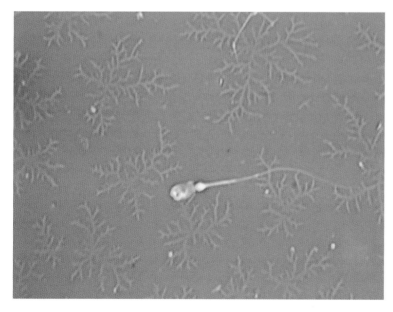

Figure 23.4 Cracking of stain due to prolonged drying. Drying can be hastened by warming of slides and by blowing warm air across the slide (eosin–nigrosin, bull, 1000×).

24

Aberrations Due to Cold Shock

During motility evaluation, observations of sperm moving backward, circling, or shimmering in place are indications of cold shock. Suspicions may be confirmed if there is a large difference between motility and live/dead assessment, or if there is increased incidence of distal midpiece reflexes without accompanying retained cytoplasmic droplets or a high incidence of bent tails.

25

Round Cells in the Ejaculate

Round cells noted in the bovine ejaculate are most commonly either immature sperm or white blood cells. Immature sperm, also known as spheroids, appear in the ejaculate following premature release from Sertoli cells and may be indicative of *immaturity, testicular degeneration, or regeneration secondary to testicular insult*. Immature sperm are quite variable in size, depending on whether the cell is a primary or a secondary spermatocyte or a spermatid [163, 164].

White blood cells in semen are often present in bulls with vesicular adenitis, ampullitis, epididymitis, or orchitis. White blood cells may also be contaminants from the prepuce or glans penis. White blood cells can be identified by visualization of the nucleus and by their uniform round shape. White blood cells with an intact membrane do not take up eosin–nigrosin stain and appear as white, somewhat irregular bodies, with a diameter of one to two times the length of a sperm head. White blood cells with broken membranes appear as pink disintegrated bodies. Positive identification and differentiation from immature sperm can be readily made by making an unstained smear, allowing it to dry and then staining it with new methylene blue, Wright's giemsa or Diff-Quik®. Once the stain is dried, the round cells can be evaluated, and final diagnosis of immature sperm cell or white blood cell can be made by the evaluator.

Figure 25.1 Round cells can be noted during evaluation of sperm motility. Staining can be used to differentiate immature sperm cells and white blood cells (wet mount, bull, 400×).

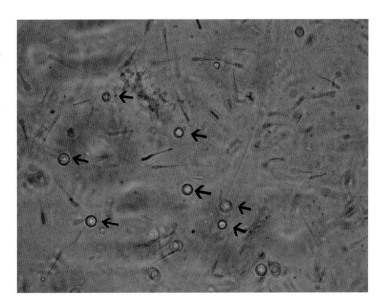

Sperm Morphology of Domestic Animals, First Edition. J.H. Koziol and C.L. Armstrong.
© 2022 John Wiley & Sons, Inc. Published 2022 by John Wiley & Sons, Inc.
Companion website: www.wiley.com/go/koziol/sperm

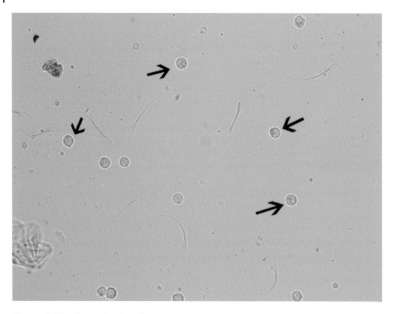

Figure 25.2   Round cells within an ejaculate (wet mount, bull, 400×).

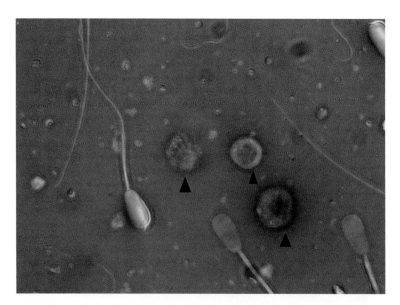

Figure 25.3   Round cells can sometimes be noted on evaluation of sperm morphology slides as noted by arrow head. Wright giemsa/Diff-Quik® staining can be used to differentiate between white blood cells and immature sperm cells (eosin–nigrosin, bull, 1000×).

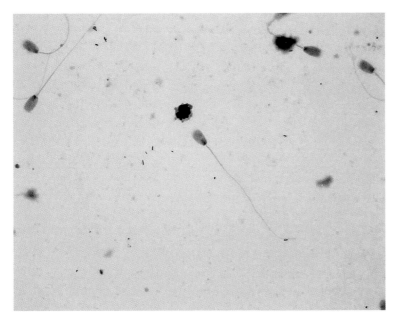

Figure 25.4 Diff-Quik®-stained slide to allow for differentiation of white blood cell from immature sperm cell. White blood cell just proximal to the sperm cell (Diff-Quik®, bull, 1000×).

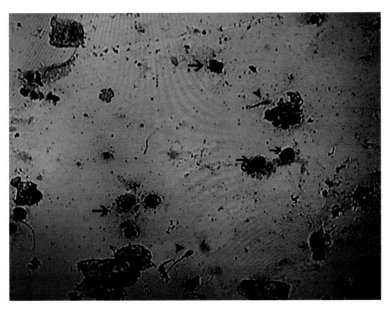

Figure 25.5 Diff-Quik®-stained slide from a bull with vesiculitis. Arrows indicate white blood cells, arrow heads indicate stained sperm (Diff-Quik®, bull, 400×).

26

Teratoids

Teratoids are severely defective sperm usually with abnormal DNA condensation and deformed nucleus, midpiece and principal pieces wrapped around the nucleus. In most cases, the tail is coiled up and lies on top of the head concealing it but other variations do occur [3]. These abnormal products of spermatogenesis may be mistaken for debris or inflammatory cells. Teratoid forms are often found at an incidence of 0–2% in normal ejaculates and 5–10% when disturbances of spermatogenesis occur [29].

Figure 26.1   Teratoid cell as noted by arrow (eosin–nigrosin, bull, 1000×).

Sperm Morphology of Domestic Animals, First Edition. J.H. Koziol and C.L. Armstrong.
© 2022 John Wiley & Sons, Inc. Published 2022 by John Wiley & Sons, Inc.
Companion website: www.wiley.com/go/koziol/sperm

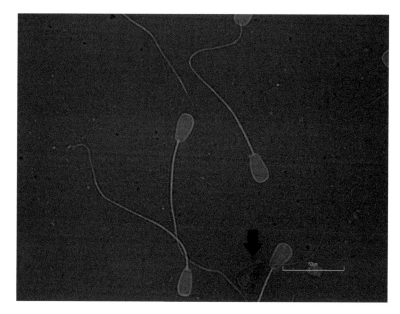

Figure 26.2 Teratoid cell as noted by arrow (eosin–nigrosin, bull 1000×).

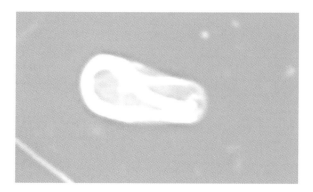

Figure 26.3 Teratoid cell (eosin–nigrosin, bull, 1000×).

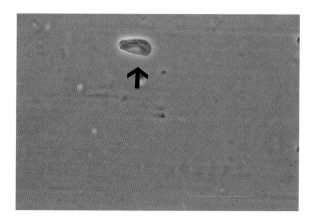

Figure 26.4 Teratoid cell (phase contrast, bull, 1000×).

27

Medussa "Cells"

Medussa formations originate from the border of ciliated epithelial cells lining the efferent ducts, which connect the rete testis to the caput epididymis. These remnants are often present in small numbers in semen smears and may increase after a severe testicular insult [149].

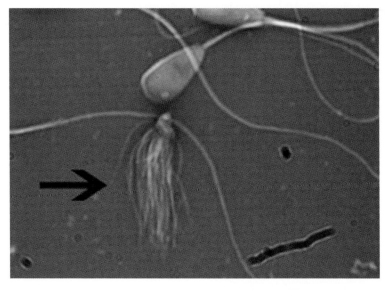

Figure 27.1 Medussa "cell" (arrow) (eosin–nigrosin, bull, 1000×).

Sperm Morphology of Domestic Animals, First Edition. J.H. Koziol and C.L. Armstrong.
© 2022 John Wiley & Sons, Inc. Published 2022 by John Wiley & Sons, Inc.
Companion website: www.wiley.com/go/koziol/sperm

28

Aberrations of Semen Quality

28.1 Oligospermia/Azoospermia

Oligospermia refers to production of ejaculates with low sperm concentrations and azoospermia refers to production of ejaculates with no sperm. When males are physically normal, failure to obtain a semen sample rarely indicates true azoospermia or oligospermia, and these diagnoses should be avoided unless there is corroborating evidence. Other means of obtaining a semen sample should be employed before this diagnosis is made, especially in farm animal species that utilize electroejaculation processes for semen collection [165] Alkaline phosphatase (AP) can be a useful biomarker in species in which the AP is derived primarily from epididymis and testicle as in the dog and the stallion [166, 167]. In the dog, AP levels >5000 U/l indicate that a complete ejaculation has occurred. If AP levels <5000 IU/l are found, then a complete ejaculate was not collected [167]. Similarly, in the stallion AP can be used to assess if a complete ejaculate was obtained. Stallions with AP concentrations over 1000 IU/l or total AP activity over 200 IU indicate a complete ejaculate [166]. In stallions, AP may be helpful in the diagnosis of ampullae blockage [166]. The presence of AP in sperm-free ejaculate is not an adequate indicator of azoospermia in the bull as it is in the dog or stallion, because only a relatively small percentage of AP within the ejaculate comes from the bull's testes and epididymis, with the majority from vesicular glands [168].

If all efforts to obtain a semen sample still indicate oligospermia or azoospermia, a genetic origin may be considered, although testicular degeneration or bilateral blockage should be ruled out. Recovery from testicular degeneration may occur after removal of the inciting cause, depending on the duration and severity of the insult. When the cause is genetic, recovery never occurs [29].

Perhaps surprisingly, ejaculates with azoospermia/oligospermia are not devoid of cells. Teratoid sperm may be found with an abundance of other cellular debris. Close inspection of what might at first be considered to be debris can reveal a high percentage of teratoid sperm.

28.2 Hemospermia

Hemospermia may be noted following collection of some males. In stallions, hemospermia has been associated with urethritis, wounds to the gland penis, idiopathic pelvic urethral rents, along with several other causes [169]. Regardless of the cause, hemorrhage is most likely from the corpus spongiosum penis (CSP). Red blood cells are associated with decreased fertility as cellular membrane integrity and motility of sperm cells can be affected [170].

Sperm Morphology of Domestic Animals, First Edition. J.H. Koziol and C.L. Armstrong.
© 2022 John Wiley & Sons, Inc. Published 2022 by John Wiley & Sons, Inc.
Companion website: www.wiley.com/go/koziol/sperm

Figure 28.1 Brown-tinged semen from a bull diagnosed with chronic vesiculitis.

If a dog is found to have hemospermia, one should suspect disease of the prostate or trauma to the glans penis or distal urethra [171].

Hemospermia is uncommon in the bull, with only one documented case in the literature. In that particular case, the source of the hemospermia was a fistula of the tip of the penis resulting in blood escaping from the corpus cavernosum during penile erection [172]. A similar case has been observed by the authors.

Bulls with penile injuries, urethral injuries, or vesiculitis sometimes present with hemospermia, but the overall case incidence is very low, consistent with the lack of reports. Bleeding from minor lacerations associated with injury during electroejaculation, penile/preputial scraping for venereal disease diagnosis, and bleeding from penile warts occurs most commonly. Clinicians should make every attempt to avoid contaminating semen samples with blood. It is reported bulls with vesiculitis have a brown-colored semen in rare cases, suggesting hemorrhage from vesicular glands (Figure 28.1). There is anecdotal evidence that fertility was lessened in bull sperm exposed to blood for a few hours [20].

28.3 Urospermia

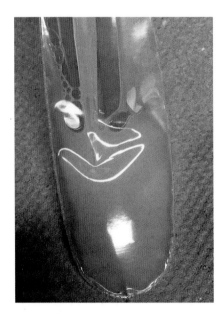

Figure 28.2 Semen sample with yellow pigment due to high riboflavin concentrations.

Occasionally, males will release urine collection into an artificial vagina or during electroejaculation. This condition is reported infrequently in the stallion and dog. In bulls, urospermia should be differentiated from semen that is yellow due to high riboflavin concentrations [173] (Figure 28.2). Urospermia has also been noted during the collection of semen by electroejaculation in both rams and bucks.

Samples containing urine are usually very dilute in contrast to ejaculates containing yellow pigments, which are typically very concentrated. Urospermia can be diagnosed if the sample has a distinct odor of urine or by the use of urea nitrogen strips (Azostix). When urine-contaminated sample is examined microscopically, sperm motility will be from minimal to absent.

In the case of ruminants, some authors advocate withholding water for several hours prior to semen collection to minimize the incidence of urospermia. However, urospermia rarely occurs during electroejaculation with proper equipment and technique, and therefore, withholding water is usually unnecessary. If urospermia does occur, a second sample can be collected almost immediately with an accelerated ramp-up phase during the ejaculation process to quickly reach maximum stimulation [174].

28.4   Antisperm Antibodies

Sperm, recognized as foreign by the body, are protected from attack from the immune system by the immunologic barriers in the testes and epididymides, which sequester sperm antigens and restrict entry of immunoglobulins and immune cells [4]. When sperm or testicular germ cells leak into the testicular or epididymal interstitium following a disruption, a marked immunogenic response to antigens contained on the sperm, culminating in autoimmunization [175]. Autoimmunization stimulates a high titer of antisperm antibodies that could persist for many years [176, 177]. Antisperm antibody formation may follow trauma, infectious and noninfectious inflammation, or by obstruction and rupture of the testicular efferent duct system, eventually leading to formation of sperm granulomas [178].

The incidence of antisperm antibodies in the bull have been reported [178–181]. In one study, the incidence of positive bulls was 35.5%; however, there was no clear evidence of any impact on fertility [178]. Furthermore, there is no correlation between breeding soundness classification or bull age and the presence of sperm agglutinating antibodies in serum or seminal fluid [181].

In the dog, antisperm antibodies have been associated with *Brucella canis* [182] and have been found to occur in males following hemicastration, epididymal aspiration, or testicular biopsy [183].

Antisperm antibodies have also been noted in the stallion in cases of infertility and following testicular trauma [184–187].

References

1 Saacke, R.G., Dejarnette, J.M., Bame, J.H. et al. (1998). Can spermatozoa with abnormal heads gain access to the ovum in artificially inseminated super- and single-ovulating cattle? *Theriogenology* 50: 117–128.

2 Johnson, W.H. (1997). The significance to bull fertility of morphologically abnormal sperm. *Vet. Clin. North Am. Food Anim. Pract.* 13: 255–270.

3 Barth, A.D. and Oko, R.J. (1989). *Abnomal Morphology of Bovine Spermatozoa*. Malden, MA: Blackwell Publishing Inc.

4 Mital, P., Hinton, B.T., and Dufour, J.M. (2011). The blood-testis and blood-epididymis barriers are more than just their tight junctions. *Biol. Reprod.* 84: 851–858.

5 Barth, A.D. and Bowman, P.A. (1994). The sequential appearance of sperm abnormalities after scrotal insulation or dexamethasone treatment in bulls. *Can. Vet. J.* 35: 93–102.

6 Welsh, T., McCraw, R., and Johnson, B. (1979). Influence of corticosteroids on testosterone production in the bull. *Biol. Reprod.* 21: 755–763.

7 Welsh, T., Randel, R., and Johnson, B. (1979). Temporal relationships among peripheral blood concentrations of corticosteroids, luteinizing hormone and testosterone in bulls. *Theriogenology* 12: 169–179.

8 Welsh, T.H. Jr. and Johnson, B.H. (1981). Stress-induced alterations in secretion of corticosteroids, progesterone, luteinizing hormone, and testosterone in bulls. *Endocrinology* 109: 185–190.

9 Knudsen, O. (1954). Cytomorphological investigations into the spermiocytogenesis of bulls with normal fertility and bulls with acquired disturbances in spermiogenesis. *Acta Pathol. Microbiol. Scand. Suppl.* 101: 1–79.

10 Senger, P.L. (2012). *Pathways to Pregnancy and Parturition*, 3e. Redmond, OR: Current Conceptions.

11 Foote, R., Swierstra, E., and Hunt, W. (1972). Spermatogenesis in the dog. *Anat. Rec.* 173: 341–351.

12 Robaire, B. and Viger, R.S. (1995). Regulation of epididymal epithelial cell functions. *Biol. Reprod.* 52: 226–236.

13 Dacheux, J.L., Gatti, J.L., and Dacheux, F. (2003). Contribution of epididymal secretory proteins for spermatozoa maturation. *Microsc. Res. Tech.* 61: 7–17.

14 Caballero, J., Frenette, G., and Sullivan, R. (2010). Post testicular sperm maturational changes in the bull: important role of the epididymosomes and prostasomes. *Vet. Med. Int.* 2011: 757194.

Sperm Morphology of Domestic Animals, First Edition. J.H. Koziol and C.L. Armstrong.
© 2022 John Wiley & Sons, Inc. Published 2022 by John Wiley & Sons, Inc.
Companion website: www.wiley.com/go/koziol/sperm

15 Amann, R.P. and Almquist, J.O. (1962). Reproductive capacity of dairy bulls. VI. Effect of unilateral vasectomy and ejaculation frequency on sperm reserves; aspects of epididymal physiology. *J. Reprod. Fertil.* 3: 260–268.

16 Linde-Forsberg, C. (2007). Biology of reproduction of the dog and modern reproductive technology. In: *The Genetics of the Dog*, 401–432. USA: CABI Publishing.

17 França, L. and Cardoso, F. (1998). Duration of spermatogenesis and sperm transit time through the epididymis in the Piau boar. *Tissue Cell* 30: 573–582.

18 Swierstra, E. (1966). Structural composition of Shorthorn bull testes and daily spermatozoa production as determined by quantitative testicular histology. *Can. J. Anim. Sci.* 46: 107–119.

19 Amann, R.P. (1970). Sperm production rates. In: *Development, Anatomy, and Physiology* (eds. A.D. Johnson, W.R. Gomes and N.L. Vandemark), 433–482. Academic Press.

20 Sekoni, V., Gustafsson, B., and Mather, E. (1980). Influence of wet fixation, staining techniques, and storage time on bull sperm morphology. *Nord. Vet. Med.* 33: 161–166.

21 Menon, A.G., Thundathil, J.C., Wilde, R. et al. (2011). Validating the assessment of bull sperm morphology by veterinary practitioners. *Can. Vet. J.* 52: 407–408.

22 Kubo-Irie, M., Matsumiya, K., Iwamoto, T. et al. (2005). Morphological abnormalities in the spermatozoa of fertile and infertile men. *Mol. Reprod. Dev* 70: 70–81.

23 Ostermeier, G.C., Sargeant, G.A., Yandell, B.S. et al. (2001). Relationship of bull fertility to sperm nuclear shape. *J. Androl.* 22: 595–603.

24 Enciso, M., Cisale, H., Johnston, S.D. et al. (2011). Major morphological sperm abnormalities in the bull are related to sperm DNA damage. *Theriogenology* 76: 23–32.

25 Wiltbank, J. and Parish, N. (1986). Pregnancy rate in cows and heifers bred to bulls selected for semen quality. *Theriogenology* 25: 779–783.

26 Saacke, R., Dalton, J., Nadir, S. et al. (2000). Relationship of seminal traits and insemination time to fertilization rate and embryo quality. *Anim. Reprod. Sci.* 60: 663–677.

27 Hancock, J. (1951). A staining technique for the study of temperature-shock in semen. *Nature* 167: 323–324.

28 Tanghe, S., Van Soom, A., Sterckx, V. et al. (2002). Assessment of different sperm quality parameters to predict in vitro fertility of bulls. *Reprod. Domest. Anim.* 37: 127–132.

29 Barth, A.D. (2013). *Bull Breeding Soundness*, 3e. Saskatoon: Western Canadian Association of Bovine Practitioners.

30 Dowsett, K., Osborne, H., and Pattie, W. (1984). Morphological characteristics of stallion spermatozoa. *Theriogenology* 22: 463–472.

31 Dowsett, K. and Knott, L. (1996). The influence of age and breed on stallion semen. *Theriogenology* 46: 397–412.

32 Love, C., Varner, D., and Thompson, J. (2000). Intra-and inter-stallion variation in sperm morphology and their relationship with fertility. *J. Reprod. Fertil. Suppl.*: 93–100.

33 Jasko, D., Lein, D., and Foote, R. (1990). Determination of the relationship between sperm morphologic classifications and fertility in stallions: 66 cases (1987–1988). *J. Am. Vet. Med. Assoc.* 197: 389–394.

34 Martínez, A.I.P. (2004). Canine fresh and cryopreserved semen evaluation. *Anim. Reprod. Sci.* 82–83: 209–224.

35 Liu, D.Y. and Baker, H.W. (1992). Morphology of spermatozoa bound to the zona pellucida of human oocytes that failed to fertilize in vitro. *J. Reprod. Fertil.* 94: 71–84.

36 Menkveld, R., Holleboom, C.A., and Rhemrev, J.P. (2011). Measurement and significance of sperm morphology. *Asian J. Androl.* 13: 59–68.

37 Walters, A.H., Eyestone, W.E., Saacke, R.G. et al. (2005). Bovine embryo development after IVF with spermatozoa having abnormal morphology. *Theriogenology* 63: 1925–1937.

38 Rathore, A.K. (1969). A note on the effect of scrotal wool cover on morphological changes in ram spermatozoa due to heat stress. *Anim. Prod.* 11: 561–563.

39 Thundathil, J., Palasz, A.T., Mapletoft, R.J. et al. (1999). An investigation of the fertilizing characteristics of pyriform-shaped bovine spermatozoa. *Anim. Reprod. Sci.* 57: 35–50.

40 Krzanowska, H. (1974). The passage of abnormal spermatozoa through the uterotubal junction of the mouse. *J. Reprod. Fertil.* 38: 81–90.

41 Nestor, A. and Handel, M.A. (1984). The transport of morphologically abnormal sperm in the female reproductive tract of mice. *Gamete Res.* 10: 119–125.

42 Saacke, R., Nebel, R., Karabinus, D., et al. (1988). Sperm transport and accessory sperm evaluation. Proc 12th Tech Con AI and Reprod, 7.

43 Kot, M.C. and Handel, M.A. (1987). Binding of morphologically abnormal sperm to mouse egg zonae pellucidae in vitro. *Gamete Res.* 18: 57–66.

44 Bane, A. and Nicander, L. (1965). Pouch formations by invaginations of the nuclear envelope of bovine and porcine sperm as a sign of disturbed spermiogenesis. *Nord. Vet. Med.* 17: 628–632.

45 Blom, E. and Birch-Andersen, A. (1965). The ultrastructure of the bull sperm. II. The sperm head. *Nord. Vet. Med.* 17: 193–212.

46 Fawcett, D.W. (1975). The mammalian spermatozoon. *Dev. Biol.* 44: 394–436.

47 Hrudka, F. (1983). Acrosomo-nuclear syndrome in canine sperm. *Andrologia* 15: 310–321.

48 Saacke, R.G. and Almquist, J.O. (1964). Ultrastructure of bovine spermatozoa. I. The head of normal, ejaculated sperm. *J. Anat.* 115: 143–161.

49 Truitt-Gibert, A.J. and Johnson, L. (1980). The crater defect in boar spermatozoa: A correlative study with transmission electron microscopy, scanning electron microscopy, and light microscopy. *Gamete Res.* 3: 259–266.

50 Bedford, J. (1964). Fine structure of the sperm head in ejaculate and uterine spermatozoa of the rabbit. *J. Reprod. Fertil.* 7: 221–228.

51 Fawcett, D.W., Anderson, W.A., and Phillips, D.M. (1971). Morphogenetic factors influencing the shape of the sperm head. *Dev. Biol.* 26: 220–251.

52 Bellvé, A.R., Anderson, E., and Hanley-Bowdoin, L. (1975). Synthesis and amino acid composition of basic proteins in mammalian sperm nuclei. *Dev. Biol.* 47: 349–365.

53 Bane, A. and Nicander, L. (1966). Electron and light microscopical studies on spermateliosis in a boar with acrosome abnormalities. *J. Reprod. Fertil.* 11: 133–138.

54 Oko, R. (1977). *Normal and Defective Bovine Spermatogenesis*. Calgary: University of Calgary.

55 Coulter, G. (1976). Effect of dexamethasone on the incidence of the "crater" defect of bovine sperm. *Theriogenology* 9: 165–173.

56 Jiráncek, E. and Rob, O. (1971). Examination of vacuoles in bull sperm nucleoplasma from fresh and deep frozen semen. *Vet. Med. (Praha)* 16: 495–500.

57 Miller, D., Hrudka, F., Cates, W. et al. (1982). Infertility in a bull with a nuclear sperm defect: a case report. *Theriogenology* 17: 611–621.

58 Barth, A. (1984). The effect of nuclear vacuoles in bovine spermatozoa on fertility in superovulated heifers. Proc Can West Soc Reprod Biol, Saskatoon, pp. 4–5.

59 Larsen, R.E. and Chenoweth, P.J. (1990). Diadem/crater defects in spermatozoa from two related Angus bulls. *Mol. Reprod. Dev* 25: 87–96.

60 Pilip, R., Del Campo, M.R., Barth, A.D. et al. (1996). In vitro fertilizing characteristics of bovine spermatozoa with multiple nuclear vacuoles: A case study. *Theriogenology* 46: 1–12.

61 Janett, F., Wild, P., and Thun, R. (1998). Nuclear vacuoles in spermatozoa: a possible cause of low fertility in the stallion. *Reprod. Domest. Anim.* 33: 399–403.

62 Thundathil, J.C. (2001). *In vitro Fertilizing Characteristics of Bovine Sperm with Abnormal Morphology*, 274. Ann Arbor: The University of Saskatchewan (Canada).

63 Held, J.P., Prater, P., and Stettler, M. (1991). Spermatozoal head defect as a cause of infertility in a stallion. *J. Am. Vet. Med. Assoc.* 199: 1760–1761.

64 Gustavsson, I. (1969). Cytogenetics, distribution and phenotypic effects of a translocation in Swedish cattle. *Hereditas* 63: 68–169.

65 Bertschinger, H.J. (1975). *The Hereditary Occurrence of Diploid Spermatozoa in the Semen of Brown Swiss bulls.* University of Zurich.

66 Blom, E. (1980). Rolled-head and nuclear-crest sperm defects in a rare case of presumed diploidy in the bull. *Nord. Vet. Med.* 32: 409–416.

67 Cran, D., Dott, H., and Wilmington, J. (1982). The structure and formation of rolled and crested bull spermatozoa. *Gamete Res.* 5: 263–269.

68 Ward, W.S. and Coffey, D.S. (1991). DNA packaging and organization in mammalian spermatozoa: comparison with somatic cells. *Biol. Reprod.* 44: 569–574.

69 Agarwal, A. and Said, T.M. (2012). Sperm chromatin assessment. In: *Textbook of Assisted Reproductive Techniques Fourth Edition: Volume 1: Laboratory Perspectives*, 75–95. CRC Press.

70 Evenson, D.P. (2016). The sperm chromatin structure assay (SCSAR®) and other sperm DNA fragmentation tests for evaluation of sperm nuclear DNA integrity as related to fertility. *Anim. Reprod. Sci.* 169: 56–75.

71 Dobrinski, I., Hughes, H.P., and Barth, A.D. (1994). Flow cytometric and microscopic evaluation and effect on fertility of abnormal chromatin condensation in bovine sperm nuclei. *J. Reprod. Fertil.* 101: 531–538.

72 Boe-Hansen, G.B., Ersbøll, A.K., Greve, T. et al. (2005). Increasing storage time of extended boar semen reduces sperm DNA integrity. *Theriogenology* 63: 2006–2019.

73 Morrell, J.M., Johannisson, A., Dalin, A.-M. et al. (2008). Sperm morphology and chromatin integrity in Swedish warmblood stallions and their relationship to pregnancy rates. *Acta Vet. Scand.* 50: 1–7.

74 Peris, S.I., Morrier, A., Dufour, M. et al. (2004). Cryopreservation of ram semen facilitates sperm DNA damage: relationship between sperm andrological parameters and the sperm chromatin structure assay. *J. Androl.* 25: 224–233.

75 Lunstra, D.-D. and Echternkamp, S. (1982). Puberty in beef bulls: acrosome morphology and semen quality in bulls of different breeds. *J. Anim. Sci.* 55: 638–648.

76 Cran, D. and Dott, H. (1976). The ultrastructure of knobbed bull spermatozoa. *J. Reprod. Fertil.* 47: 407–408.

77 Hurtgen, J. and Johnson, L. (1982). Fertility of stallions with abnormalities of the sperm acrosome. *J. Reprod. Fertil. Suppl.* 32: 15–20.

78 Revell, S. and Chasey, D. (1988). Morphological defects of the acrosome in boar spermatozoa. *Res. Vet. Sci.* 45: 149–151.

79 Ott, R., Heath, E., and Bane, A. (1982). Abnormal spermatozoa, testicular degeneration, and varicocele in a ram. *Am. J. Vet. Res.* 43: 241–245.

80 Santos, N.R., Krekeler, N., Schramme-Jossen, A. et al. (2006). The knobbed acrosome defect in four closely related dogs. *Theriogenology* 66: 1626–1628.

81 Oettle, E. and Soley, J. (1988). Abnormalities in spermatozoa from dogs: a light and electron microscopic study. *Vet. Med. Nachr.* 59: 28–70.

82 Barth, A.D. (1986). The knobbed acrosome defect in beef bulls. *Can. Vet. J.* 27: 379–384.

83 Donald, H. and Hancock, J. (1953). Evidence of gene-controlled sterility in bulls. *J. Agric. Sci.* 43: 178–181.

84 Chenoweth, P.J. (2005). Genetic sperm defects. *Theriogenology* 64: 457–468.

85 Thundathil, J., Meyer, R., Palasz, A.T. et al. (2000). Effect of the knobbed acrosome defect in bovine sperm on IVF and embryo production. *Theriogenology* 54: 921–934.

86 Thundathil, J., Palasz, A.T., Barth, A.D. et al. (2002). Plasma membrane and acrosomal integrity in bovine spermatozoa with the knobbed acrosome defect. *Theriogenology* 58: 87–102.

87 Thundathil, J., Palomino, J., Barth, A. et al. (2001). Fertilizing characteristics of bovine sperm with flattened or indented acrosomes. *Anim. Reprod. Sci.* 67: 231–243.

88 Meyer, R.A. and Barth, A.D. (2001). Effect of acrosomal defects on fertility of bulls used in artificial insemination and natural breeding. *Can. Vet. J.* 42: 627.

89 Cooper, A.M. and Peet, R.L. (1983). Infertility in a Hereford bull associated with increased numbers of detached sperm heads in his ejaculate. *Aust. Vet. J.* 60: 225–226.

90 Thilander, G., Settergren, I., and Plöen, L. (1985). Abnormalities of testicular origin in the neck region of bull spermatozoa. *Anim. Reprod. Sci.* 8: 151–157.

91 Barth, A.D. (2007). Sperm accumulation in the ampullae and cauda epididymides of bulls. *Anim. Reprod. Sci.* 102: 238–246.

92 Lino, B., Braden, A., and Turnbull, K. (1967). Fate of unejaculated spermatozoa. *Nature* 213: 594–595.

93 Holtz, W. and Foote, R. (1972). Sperm production, output and urinary loss in the rabbit. *Proc. Exp. Biol. Med.* 141: 958–962.

94 McCue, P.M., Scoggin, C.F., Moffett, P.D. et al. (2014). Spermiostasis in stallions: a retrospective study of clinical cases. *J. Equine Vet. Sci.* 34: 47.

95 Câmara, L., Câmara, D., Maiorino, F. et al. (2018). Canine testicular disorders and their influence on sperm morphology. *Anim. Reprod.* 11: 32–36.

96 Ortega-Pacheco, A., Rodríguez-Buenfil, J., Segura-Correa, J. et al. (2006). Pathological conditions of the reproductive organs of male stray dogs in the tropics: prevalence, risk factors, morphological findings and testosterone concentrations. *Reprod. Domest. Anim.* 41: 429–437.

97 Williams, G. (1965). An abnormality of the spermatozoa of some Hereford bulls. *Vet. Rec.* 77: 1204–1206.

98 Hancock, J. and Rollinson, D. (1949). A seminal defect associated with sterility of Guernsey bulls. *Vet. Rec.* 61: 742.

99 Bloom, G. and Nicander, L. (1961). On the ultrastructure and development of the protoplasmic droplet of spermatozoa. *Cell Tissue Res.* 55: 833–844.

100 Jones, W. (1962). Abnormal morphology of the spermatozoa in Guernsey bulls. *Brit. Vet. J.* 118: 257.

101 Blom, E. (1977). A decapitated sperm defect in two sterile Hereford bulls. *Nord. Vet. Med.* 29: 119–123.

102 Pesch, S. and Bergmann, M. (2006). Structure of mammalian spermatozoa in respect to viability, fertility and cryopreservation. *Micron.* 37: 597–612.

103 Morton, D.B. and Bruce, S.G. (1989). Semen evaluation, cryopreservation and factors relevant to the use of frozen semen in dogs. *J. Reprod. Fertil. Suppl.* 39: 311–316.

104 Johnson, K.R., Dewey, C.E., Bobo, J.K. et al. (1998). Prevalence of morphologic defects in spermatozoa from beef bulls. *J. Am. Vet. Med. Assoc.* 213: 1468–1471.

105 Lagerlöf, N. (1934). Morphological studies on the change in sperm structure and in the testes of bulls with decreased or abolished fertility. *Acta Pathol. Microbiol. Scand.* 19: 254–266.

106 Ruttle, J., Ezaz, Z., and Sceery, E. (1975). Some factors influencing semen characteristics in range bulls. *J. Anim. Sci.* 41: 1069–1076.

107 Saacke, R. (1970). Morphology of the sperm and its relationship to fertility. *Proc Third Technical Conference on Artificial Insemination and Reproduction*; 17.

108 Smith, M.F., Morris, D.L., Amoss, M.S. et al. (1981). Relationships among fertility, scrotal circumference, seminal quality, and libido in Santa Gertrudis bulls. *Theriogenology* 16: 379–397.

109 Blom, E. Sperm morphology with reference to bull infertility. *Proc of the First All-India Symp Anim Reprod, Ludhiana* 1977;61-81.

110 Dott, H. and Dingle, J. (1968). Distribution of lysosomal enzymes in the spermatozoa and cytoplasmic droplets of bull and ram. *Exp. Cell. Res.* 52: 523–540.

111 Söderquist, L., Janson, L., Larsson, K. et al. (1991). Sperm morphology and fertility in AI bulls. *Transbound. Emerg. Dis.* 38: 534–543.

112 Mortimer, R., Seidel, G., Amann, R. et al. (1991). Use of in vitro fertilization to evaluate spermatozoa with proximal droplets in young beef bulls. *Theriogenology* 35: 247.

113 Amann, R., Seidel, G., and Mortimer, R. (2000). Fertilizing potential in vitro of semen from young beef bulls containing a high or low percentage of sperm with a proximal droplet. *Theriogenology* 54: 1499–1515.

114 Peña, A.I., Barrio, M., Becerra, J.J. et al. (2007). Infertility in a dog due to proximal cytoplasmic droplets in the ejaculate: investigation of the significance for sperm functionality in vitro. *Reprod. Domest. Anim.* 42: 471–478.

115 Blom, E. and Birch-Andersen A (1968) The ultrastructure of the "pseudo-droplet-defect" in bull sperm. *VIth Congr Reprod Insem Artif* :117.

116 Chenoweth, P., Chase, C., Risco, C. et al. (2000). Characterization of gossypol-induced sperm abnormalities in bulls. *Theriogenology* 53: 1193–1203.

117 Brito, L.F.C. (2007). Evaluation of stallion sperm morphology. *Clin. Tech. Equine Pract.* 6: 249–264.

118 Heath, E., Aire, T., and Fujiwara, K. (1985). Microtubular mass defect of spermatozoa in the stallion. *Am. J. Vet. Res.* 46: 1121–1125.

119 Chenoweth, P., McDougall, H.L., McCosker, P. et al. (1970). An abnormality of the spermatozoa of a stallion (Equus caballus). *Brit. Vet. J.* 126: 476–481.

120 Brito, L., Kelleman, A., Greene, L. et al. (2010). Semen characteristics in a sub-fertile Arabian stallion with idiopathic teratospermia. *Reprod. Domest. Anim.* 45: 354–358.

121 Bonet, S., Briz, M., and Fradera, A. (1993). Ultrastructural abnormalities of boar spermatozoa. *Theriogenology* 40: 383–396.

122 Blom, E. (1959). A rare sperm abnormality: 'corkscrew-sperms' associated with sterility in bulls. *Nature* 183: 1280–1281.

123 Blom, E. (1978). The corkscrew sperm defect in Danish bulls - a possible indicator of nuclear fallout? *Nord. Vet. Med.* 30: 1–8.

124 Blom, E. (1966). A new sterilizing and hereditary defect (the 'Dag defect') located in the bull sperm tail. *Nature* 209: 739–740.

125 Hellmen, E., Ploen, L., Settergren, I. et al. (1980). Middle piece defects of testicular origin in bull sperm. *Nord. Vet. Med.* 32: 423–426.

126 Wenkoff, M.S. (1978). A sperm mid-piece defect of epididymal origin in two Hereford bulls. *Theriogenology* 10: 275–281.

127 Koefoed-Johnsen, H., Andersen, J., Andresen, E. et al. (1980). The Dag defect of the tail of the bull sperm. Studies on the inheritance and pathogenesis. *Theriogenology* 14: 471–475.

128 Koefoed-Johnsen, H.H. and Pedersen, H. (1971). Further observations on the Dag-defect of the tail of the bull spermatozoon. *J. Reprod. Fertil.* 26: 77–83.

129 Hellander, J.C., Samper, J.C., and Crabo, B.G. (1991). Fertility of a stallion with low sperm motility and a high incidence of an unusual sperm tail defect. *Vet. Rec.* 128: 449–451.

130 Rota, A., Manuali, E., Caire, S. et al. (2008). Severe tail defects in the spermatozoa ejaculated by an English bulldog. *J. Vet. Med. Sci.* 70: 123–125.

131 Goyal, H. (1983). Histoquantitative effects of orchiectomy with and without testosterone enanthate treatment on the bovine epididymis. *Am. J. Vet. Res.* 44: 1085–1090.

132 Cupps, P., Laben, R., and Rahlmann, D. (1960). Effects of estradiol benzoate injections on the characteristics of bovine semen. *J. Dairy Sci.* 43: 1135–1139.

133 Cupps, P.T. and Briggs, J. (1965). Changes in the epididymis associated with morphological changes in the spermatozoa. *J. Dairy Sci.* 48: 1241–1244.

134 Swanson, E. and Boatman, J. (1953). The effect of thiouracil feeding upon the seminal characteristics of dairy bulls[1]. *J. Dairy Sci.* 36: 246–252.

135 Burgess, G. and Chenoweth, P. (1975). Mid piece abnormalities in bovine semen following experimental and natural cases of bovine ephemeral fever. *Brit. Vet. J* 131: 536–544.

136 Prabhakar, J., Chimbombi, J., Malmgren, L. et al. (1990). Effects on testosterone and LH concentrations of induced testicular degeneration in bulls. *Acta Vet. Scand.* 31: 505–507.

137 Swanson, E.W. and Boyd, L.J. (1962). Factors affecting coiled-tail spermatozoa in the bull. *Am. J. Vet. Res.* 23: 300–309.

138 Barth, A.D. and Waldner, C.L. (2002). Factors affecting breeding soundness classification of beef bulls examined at the Western College of Veterinary Medicine. *Can. Vet. J.* 43: 274–284.

139 Barth, A. (1991). The pathogenesis of abnormal sperm production. *Proc of the Bull Mgmt Fertil Conf*. Manhattan, KS.

140 Barth, A. (1997). Evaluation of potential breeding soundness in the bull. In: *Current Therapy in Large Animal Theriogenology* (ed. R.S. Youngquist), 222–236. Philadelphia: WB Saunders.

141 Al-Makhzoomi, A., Lundeheim, N., Haard, M. et al. (2008). Sperm morphology and fertility of progeny-tested AI dairy bulls in Sweden. *Theriogenology* 70: 682–691.

142 Kashoma, I., Luziga, C., and Mgongo, F. (2009). Prevalence of spermatozoa morphologic defects from Zebu bulls under free mating system. *Tanz. Vet. J.* 26: 14.

143 Matoušek, J. (1980). The haemolytic factor (phospholipid-binding protein) of the bull reproductive tract—its synthesis and effect on spermatozoal cytoplasm droplets. *Anim. Reprod. Sci.* 3: 195–205.

144 Matousek, J. and Kysilka, C. (1984). The phospholipid-binding protein of the reproductive tract of the bull - effect on the removal of spermatozoal cytoplasm droplets in other species and influence of antibodies on its reactivity. *Anim. Reprod. Sci.* 7: 433–440.

145 Waberski, D., Meding, S., Dirksen, G. et al. (1994). Fertility of long-term-stored boar semen: influence of extender (Androhep and Kiev), storage time and plasma droplets in the semen. *Anim. Reprod. Sci.* 36: 145–151.

146 Chłopik, A. and Wysokińska, A. (2020). Canine spermatozoa—what do we know about their morphology and physiology? An overview. *Reprod. Domest. Anim.* 55: 113–126.

147 Luvoni, G. and Morselli, M. (2017). Canine epididymal spermatozoa: a hidden treasure with great potential. *Reprod. Domest. Anim.* 52: 197–201.

148 Bielanski, E. (1981). Bibliography on spermatozoan morphology in the stallion. *Bibliograph. Reprod.* 38: 501–596.

149 Roberts, S.J. (1971). *Veterinary Obstetrics and Genital Diseases (Theriogenology)*, 2e, 717. Ann Arbor, MI: Edward Brothers Inc.

150 Barth, A.D. (1989). Abaxial tail attachment of bovine spermatozoa and its effect on fertility. *Can. Vet. J.* 30: 656–662.

151 Blom, E. (1976). A sterilizing tail stump sperm defect in a Holstein-Friesian bull. *Nord. Vet. Med.* 28: 295–298.

152 Coubrough, R. and Barker, C. (1964). Spermatozoa: an unusual middle piece abnormality associated with sterility in bulls. *Proc. 5th Int. Congr. Anim. Reprod. (Trento)* 5: 219–229.

153 Blom, E. and Birch-Andersen, A. (1980). Ultrastructure of the tail stump sperm defect in the bull. *Acta Pathol. Microbiol. Scand. A* 88: 397–405.

154 Kawakami, E., Ozawa, T., Hirano, T. et al. (2005). Formation of detached tail and coiled tail of sperm in a Beagle dog. *J. Vet. Med. Sci.* 67: 83–85.

155 Sathananthan, A., Ratnam, S., Ng, S. et al. (1996). The sperm centriole: its inheritance, replication and perpetuation in early human embryos. *Hum. Reprod.* 11: 345–356.

156 Rawe, V., Terada, Y., Nakamura, S. et al. (2002). A pathology of the sperm centriole responsible for defective sperm aster formation, syngamy and cleavage. *Hum. Reprod.* 17: 2344–2349.

157 Navara, C., First, N., and Schatten, G. (1994). Microtubule organization in the cow during fertilization, polyspermy, parthenogenesis, and nuclear transfer: the role of the sperm aster. *Dev. Biol.* 162: 29–40.

158 Long, C.R., Pinto-Correia, C., Duby, R.T. et al. (1993). Chromatin and microtubule morphology during the first cell cycle in bovine zygotes. *Mol. Reprod. Dev* 36: 23–32.

159 Siqueira, J., Pinho, R., Guimarães, S. et al. (2010). Immotile short-tail sperm defect in Nelore (Bos taurus indicus) breed bulls. *Reprod. Domest. Anim.* 45: 1122–1125.

160 Andersson, M., Peltoniemi, O., Makinen, A. et al. (2000). The hereditary 'short tail' sperm defect-a new reproductive problem in Yorkshire boars. *Reprod. Domest. Anim.* 35: 59–63.

161 Sironen, A.I., Andersson, M., Uimari, P. et al. (2002). Mapping of an immotile short tail sperm defect in the Finnish Yorkshire on porcine chromosome 16. *Mamm. Genome* 13: 45–49.

162 Kopp, C., Sironen, A., Ijäs, R. et al. (2008). Infertile boars with knobbed and immotile short-tail sperm defects in the Finnish Yorkshire breed. *Reprod. Domest. Anim.* 43: 690–695.

163 Swerczek, T. (1975). Immature germ cells in the semen of thoroughbred stallions. *J. Reprod. Fertil. Suppl.* 23: 135–137.

164 Patil, P.S., Humbarwadi, R.S., Patil, A.D. et al. (2013). Immature germ cells in semen – correlation with total sperm count and sperm motility. *J. Cytol.* 30: 185–189.

165 Ott, R.S. (1986). Breeding soundness examination of bulls. In: *Current Therapy in Theriogenology*, 2e (ed. D.A. Morrow), 125–136. Philadelphia: W.B. Saunders Company.

166 Turner, R.M.O. and McDonnell, S.M. (2003). Alkaline phosphatase in stallion semen: characterization and clinical applications. *Theriogenology* 60: 1–10.

167 Kutzler, M.A., Solter, P.F., Hoffman, W.E. et al. (2003). Characterization and localization of alkaline phosphatase in canine seminal plasma and gonadal tissues. *Theriogenology* 60: 299–306.

168 Alexander, F.C.M., Zemjanis, R., Graham, E.F. et al. (1971). Semen characteristics and chemistry from bulls before and after seminal vesiculectomy and after vasectomy. *J. Dairy Sci.* 54: 1530–1535.

169 Brinsko, S.P., Blanchard, T.L., Varner, D.D. et al. (2011). Surgery of the stallion reproductive tract. In: *Manual of Equine Reproduction*, 3e (eds. S.P. Brinsko, T.L. Blanchard, D.D. Varner, et al.), 242–275. Saint Louis: Mosby.

170 Möller, G., Azevedo, L., Trein, C. et al. (2005). Effects of hemospermia on seminal quality. *Anim. Reprod. Sci.* 89: 264–267.

171 Memon, M. (2007). Common causes of male dog infertility. *Theriogenology* 68: 322–328.

172 Leipnitz, G. (1980). Fetilization capacity of bull semen containing blood. Befruchtungsfahigkeit von bluthaltigem Rindersperma. *Zuchthygiene* 15: 28–29.

173 White, I.G. and Lincoln, G.J. (1960). The yellow pigmentation of bull semen and its content of riboflavin, niacin, thiamine and related compounds. *Biochem. J* 76: 301–306.

174 Kastelic, J.P., Thundathil, J., and Brito, L.F. (2012). Bull BSE and semen analysis for predicting bull fertility. *Clin. Therio.* 4: 277–287.

175 Marshburn, P.B. and Kutteh, W.H. (1994). The role of antisperm antibodies in infertility. *Fertil. Steril.* 61: 799–811.

176 Purvis, K. and Christiansen, E. (1993). Infection in the male reproductive tract. Impact, diagnosis and treatment in relation to male infertility. *Int. J. Androl.* 16: 1–13.

177 Hinting, A., Soebadi, D., and Santoso, R. (1996). Evaluation of the immunological cause of male infertility. *Andrologia* 28: 123–126.

178 Zralý, Z., Bendova, J., Diblíková, I. et al. (2002). Antisperm antibodies in blood sera of bulls and correlations with age, breed and ejaculate quality. *Acta Vet. Brno* 71: 303–308.

179 Kim, C.A., Parrish, J.J., Momont, H.W. et al. (1999). Effects of experimentally generated bull antisperm antibodies on in vitro fertilization. *Biol. Reprod.* 60: 1285–1291.

180 Perez, T. and Carrasco, L. (1964). Auto-immunization against sperm as a cause of bull infertility. *Int. Cong. Anim. Reprod.* 5: 247–252.

181 Purswell, B., Dawe, D., Caudle, A. et al. (1983). Spermagglutinins in serum and seminal fluid of bulls and their relationship to fertility classification. *Theriogenology* 20: 375–381.

182 George, L. and Carmichael, L. (1984). Antisperm responses in male dogs with chronic Brucella canis infections. *Am. J. Vet. Res.* 45: 274–281.

183 Attia, K.A., Zaki, A.A., Eilts, B.E. et al. (2000). Anti-sperm antibodies and seminal characteristics after testicular biopsy or epididymal aspiration in dogs. *Theriogenology* 53: 1355–1363.

184 Zhang, J., Ricketts, S., and Tanner, S. (1990). Antisperm antibodies in the semen of a stallion following testicular trauma. *Equine Vet. J.* 22: 138–141.

185 Papa, F.O., Alvarenga, M.A., Lopes, M.D. et al. (1990). Infertility of autoimmune origin in a stallion. *Equine Vet. J.* 22: 145–146.

186 Kenney, R., Cummings, M., Teuscher, C. et al. (2000). Possible role of autoimmunity to spermatozoa in idiopathic infertility of stallions. *J. Reprod. Fertil. Suppl.*: 23–30.

187 Day, M. (1996). Detection of equine antisperm antibodies by indirect immunofluorescence and the tube-slide agglutination test. *Equine Vet. J.* 28: 494–496.

Index

Sperm Morphology of Domestic Animals, First Edition. J.H. Koziol and C.L. Armstrong.
© 2022 John Wiley & Sons, Inc. Published 2022 by John Wiley & Sons, Inc.
Companion website: www.wiley.com/go/koziol/sperm